Coil Tubing Unit for Oil Production and Remedial Measures

RIVER PUBLISHERS SERIES IN CHEMICAL, ENVIRONMENTAL, AND ENERGY ENGINEERING

Series Editors:

MEDANI P. BHANDARI
Akamai University, USA; Summy State University, Ukraine and Atlantic State Legal Foundation, NY, USA

JACEK BINDA
PhD, Rector of the International Affairs, Bielsko-Biala School of Finance and Law, Poland

DURGA D. POUDEL
PhD, University of Louisiana at Lafayette, Louisiana, USA

SCOTT GARNER
JD, MTax, MBA, CPA, Asia Environmental Holdings Group (Asia ENV Group), Asia Environmental Daily, Beijing/Hong Kong, People's Republic of China

HANNA SHVINDINA
Sumy State University, Ukraine
and
ALIREZA BAZARGAN
NVCo and University of Tehran, Iran

The "River Publishers Series in Chemical, Environmental, and Energy Engineering" is a series of comprehensive academic and professional books which focus on Environmental and Energy Engineering subjects. The series focuses on topics ranging from theory to policy and technology to applications.

Books published in the series include research monographs, edited volumes, handbooks and textbooks. The books provide professionals, researchers, educators, and advanced students in the field with an invaluable insight into the latest research and developments.

Topics covered in the series include, but are by no means restricted to the following:

- Energy and Energy Policy
- Chemical Engineering
- Water Management
- Sustainable Development
- Climate Change Mitigation
- Environmental Engineering
- Environmental System Monitoring and Analysis
- Sustainability: Greening the World Economy

For a list of other books in this series, visit www.riverpublishers.com

Coil Tubing Unit for Oil Production and Remedial Measures

Mohammed Ismail Iqbal

University of Technology and Applied Sciences, Nizwa, Oman

LONDON AND NEW YORK

Published 2021 by River Publishers
River Publishers
Alsbjergvej 10, 9260 Gistrup, Denmark
www.riverpublishers.com

Distributed exclusively by Routledge
4 Park Square, Milton Park, Abingdon, Oxon OX14 4RN
605 Third Avenue, New York, NY 10017, USA

Coil Tubing Unit for Oil Production and Remedial Measures / by Mohammed Ismail Iqbal.

© 2021 River Publishers. All rights reserved. No part of this publication may be reproduced, stored in a retrieval systems, or transmitted in any form or by any means, mechanical, photocopying, recording or otherwise, without prior written permission of the publishers.

Routledge is an imprint of the Taylor & Francis Group, an informa business

ISBN 978-87-7022-690-5 (print)

While every effort is made to provide dependable information, the publisher, authors, and editors cannot be held responsible for any errors or omissions.

Contents

Preface xiii

List of Figures xv

List of Tables xvii

List of Abbreviations xix

1 **Nitrogen Application** 1
  1.1 Introduction 1
  1.2 History of N_2 2
      1.2.1 N_2 Properties 3
  1.3 Cryogenics 3
      1.3.1 Introduction 3
  1.4 Basic Equipment 5
      1.4.1 Storage Tank 7
      1.4.2 Pumping System 7
      1.4.3 Vaporizer System 7
  1.5 Safety 8
      1.5.1 General Information 8
      1.5.2 Safety Bulletin from CGA (Compressed Gas Association) 9
      1.5.3 Oxygen-deficient Atmospheres 10
      1.5.4 Safety for Handling and Exposure 11
  1.6 N_2 Service Applications 11
      1.6.1 Displacement 12
      1.6.2 Nitrified Fluids-Acidisation 15
      1.6.3 Atomized Atom 16
      1.6.4 Foamed Acid 17
          1.6.4.1 N_2 Retention 17
          1.6.4.2 Diverting 18

		1.6.4.3	Production of Fines	18
		1.6.4.4	Foamed Acid Guidelines	19
	1.6.5	Aerating Conventional Fluids		19
	1.6.6	Pipeline Purging		20
	1.6.7	Use of Foam as a Drilling and Workover Fluid		20
1.7	Foam Clean Out			22
	1.7.1	Introduction		22
	1.7.2	Foam Stability and Viscosity		23
	1.7.3	Fire Control		24
1.8	Water Control Technique by N_2 Injection			24
	1.8.1	Introduction		24
	1.8.2	Technology		25
	1.8.3	Job Description		26
	1.8.4	Commercial Viability		26
	1.8.5	Quick and Easy		26
	1.8.6	Versatility and Adaptability		26
	1.8.7	Economical		26
	1.8.8	Freeding Differentially Stuck Drill Pipe		27
		1.8.8.1	N_2 Lift	28
		1.8.8.2	N_2 cushion	28
1.9	Case Study - I			28
1.10	Results/Remarks			29
1.11	Conclusion			30
1.12	Specification of N_2 Pumpers Available with WSS COLD END			30

2 Water Control 31

2.1	Introduction to Water Production			31
	2.1.1	Methods to Predict, Prevent, Delay and Reduce Excessive Water Production		32
		2.1.1.1	Oil and Water production rates and ratios	32
			2.1.1.1.1 Material Mass Balance	32
			2.1.1.1.2 Darcy's Law	33
			2.1.1.1.3 Productivity index	34
			2.1.1.1.4 Simulators	34
		2.1.1.2	Rate-limited facilities	34
		2.1.1.3	Water production effect on bypassed oil	35
		2.1.1.4	Reservoir maturity	35

		2.1.1.5	Water production rate effect on corrosion rates	36
		2.1.1.6	Water production rate effect on scale deposition rates	36
		2.1.1.7	Water production rate effect on sand production	36
2.2	Water Production Mechanisms			37
	2.2.1	Completions-Related Mechanisms		37
		2.2.1.1	Casing leaks	37
		2.2.1.2	Channel behind casing	38
		2.2.1.3	Completion into Water	38
	2.2.2	Reservoir-Related Mechanisms		38
		2.2.2.1	Bottomwater	38
		2.2.2.2	Barrier breakdown	38
		2.2.2.3	Coning and cresting	39
		2.2.2.4	Channeling through high permeability . .	39
		2.2.2.5	Fracture communication between injector and producer	40
		2.2.2.6	Stimulation out of zone	41
2.3	Preventing Excessive Water Production			41
	2.3.1	Preventing Casing Leaks		41
	2.3.2	Preventing Channels Behind Casing		41
	2.3.3	Preventing Coning and Cresting		42
	2.3.4	Perforating		42
	2.3.5	Fracturing		43
	2.3.6	Artificial Barriers		43
	2.3.7	Dual Completions		43
	2.3.8	Horizontal Wells to Prevent Coning		43
	2.3.9	Preventing Channeling Through High Permeability .		44
		2.3.9.1	Perforating	45
		2.3.9.2	Stimulation techniques	45
		2.3.9.3	Permeability reduction	46
		2.3.9.4	Preventing fracture communication between injector and producer	47
		2.3.9.5	Completing to accommodate future water production rates future zonal isolation . .	48
2.4	Creative Water Management			49
2.5	Treatments Used to Reduce Excessive Water Production . .			51
	2.5.1	Characterizing the Problem		51

	2.5.2	Treatment Design		52
	2.5.3	Expected Treatment Effect on Water Production		52
	2.5.4	Treatment Types		53
		2.5.4.1	Zone sealants	53
		2.5.4.2	Permeability-Reducing Agents (PRA)	54
		2.5.4.3	Relative Permeability Modifiers (RPM)	55
	2.5.5	Description of Previously Applied Treatments		56
		2.5.5.1	Mechanical plugs	56
		2.5.5.2	Sand plugs	58
		2.5.5.3	Water-based cement	58
		2.5.5.4	Hydrocarbon-based cements	58
		2.5.5.5	Externally activated silicates	58
		2.5.5.6	Internally Activated Silicates (IAS)	58
		2.5.5.7	Monomer systems	59
		2.5.5.8	Crosslinked polymer systems	59
		2.5.5.9	Surface-active RPMs	60
		2.5.5.10	Foams	60
	2.5.6	Treatment Lifetime		60
2.6	Selecting Treatment Composition and Volume			62
	2.6.1	Placement Techniques		63
		2.6.1.1	Bullheading	63
		2.6.1.2	Mechanical packer placement	63
		2.6.1.3	Dual injection	63
		2.6.1.4	Isoflow	63
	2.6.2	Viscosity Considerations		64
	2.6.3	Temperature Considerations		64
	References			65

3 Sand Control — **69**

3.1	Sand Control Introduction			69
	3.1.1	Formation Damage		69
	3.1.2	Fines Migration		70
	3.1.3	Sand Production Mechanisms		71
3.2	Formation Sand			72
	3.2.1	Petro Physical Properties		73
	3.2.2	Geological Deposition of Sand		73
		3.2.2.1	Desert aeolian sands	73
		3.2.2.2	Marine shelf sand	73
		3.2.2.3	Beaches, barriers and bar	74

		3.2.2.4	Tidal flat and estuarine sands	74
		3.2.2.5	Fluviatile sands	74
		3.2.2.6	Alluvial sands	74
	3.2.3	Formation Sand Description		74
		3.2.3.1	Quicksand	74
		3.2.3.2	Partially consolidated sand	75
		3.2.3.3	Friable sand	75
3.3	Causes and Effects of Sand Production			75
	3.3.1	Causes of Sand Production		75
		3.3.1.1	Totally unconsolidated formation	75
		3.3.1.2	High production rates	75
		3.3.1.3	Water productions	76
		3.3.1.4	Increase in water production	76
		3.3.1.5	Reservoir depletion	76
	3.3.2	Effects of Sand Production		76
3.4	Detection and Prediction of Sand Production			77
	3.4.1	Methods for Monitoring and Detection of Sand Production		77
		3.4.1.1	Wellhead shakeouts	77
		3.4.1.2	Safety plugs and erosion sand probes . . .	77
		3.4.1.3	Sonic sand detection	77
3.5	Methods for Sand Exclusion			78
	3.5.1	Production Restriction		78
	3.5.2	Mechanical Methods		79
	3.5.3	In-Situ Chemical Consolidation Methods		79
	3.5.4	Combination Methods		79
	3.5.5	Selecting the Appropriate Sand Exclusion Method .		80
3.6	Mechanical Methods of Sand Exclusions			80
	3.6.1	Mechanical Components		83
		3.6.1.1	Pack-sands	83
		3.6.1.2	Liners and screens	83
		3.6.1.3	Carrier fluids	86
	3.6.2	Tools and Accessories		89
	3.6.3	Completion Tools		89
		3.6.3.1	Gravel-pack Packer	89
		3.6.3.2	Flow sub	89
		3.6.3.3	Mechanical fluid-loss device	89
		3.6.3.4	Safety joint	90
		3.6.3.5	Blank pipe	90

		3.6.3.6	Tell-tale screen	90
		3.6.3.7	Seal assembly	90
		3.6.3.8	Sump packer	91
	3.6.4	Service Tools		91
		3.6.4.1	Crossover service tool	91
		3.6.4.2	Reverse-ball check-valve	91
		3.6.4.3	Swivel joint	91
		3.6.4.4	Washpipe	91
		3.6.4.5	Shifting tools	92
		3.6.4.6	Tool selection	92
3.7	Mechanical Method: Techniques and Procedures			92
	3.7.1	Gravity Pack		93
	3.7.2	Washdown Method		93
	3.7.3	Circulation Packs		93
	3.7.4	Reverse-circulation Pack		94
	3.7.5	Bullhead Pressure Packs		94
	3.7.6	Circulating-pressure Packs		94
	3.7.7	Slurry Packs		95
	3.7.8	Staged Prepacks and Acid Prepacks		96
	3.7.9	Water-packs and High-rate Water-packs		96
	3.7.10	Fracpacks		97
	3.7.11	Summary		97
	3.7.12	Mechanical Job Designs		98
		3.7.12.1	Formation characteristics	98
		3.7.12.2	Pack-sand selection criteria	100
		3.7.12.3	Screen selection criteria	102
		3.7.12.4	Gravel-pack job calculations	106
			3.7.12.4.1 Pack-sand volume required	106
			3.7.12.4.2 Carrier-fluid Volume	109
		3.7.12.5	Predicting job outcome by computer modeling	110
3.8	Chemical Consolidation Techniques			111
	3.8.1	Internally Activated Systems		114
	3.8.2	Externally Activated Systems		114
	3.8.3	Application		115
3.9	Combination Methods			116
	3.9.1	Semicured Resin-coated Pack Gravels		116
	3.9.2	Liquid Resin-coated Pack Gravel		117
3.10	Horizontal Gravel-Packing			119

3.10.1	Variables that Affect Sand Delivery	121
3.10.2	Pump Rate and Fluid Velocity	121
3.10.3	Alpha and Beta Wave Progression Through the Annulus	122
3.10.4	Sand Concentration	123
3.10.5	Placement Procedure and Tool Configuration	124
3.10.6	Liner/Tailpipe Ratio	125
3.10.7	Screen/Casing Clearance	125
3.10.8	Perforation Phasing	126
References		126

Index   129

About the Author   133

Preface

Good production and reservoir engineers look for opportunities in improving well performance irrespective of the various problems associated. When the production of fluid take place, there are various losses that takes place (across reservoir, across completion which is largely related to perforation, across tubing, sub surface safety valve, etc.). The magnitude of these individual pressure losses depends on the reservoir properties and pressures, fluid produced and the well design. Production engineers need to understand the interplay of these various factors, so that maximum profitability from oil and gas well/reservoir takes place.

During the initial stage of production, water production may not be in large quantities but as the reservoir drops down below the bubble point pressure, larger production of water tends to come from crude, due to which the stage of production does not become economical because of which many issues like relative expense of completing wells to main low water production rates, water production effect on corrosion rates, sand production, scale formation, and so on takes place. This book describes the various mechanisms to prevent excessive water production through casing leaks, behind casing, preventing conning and cresting, perforating, fracturing and completion are widely discussed with its treatment and management.

Another problem very often is sand control in unconsolidated reservoirs which is a costly operational problem which has a significant impact on the case of well operation and the economics of oil or gas production. It is usually associated with shallow, young formation, but has also been encountered at greater depths too. This book describes sources of sand production, detection and prediction of sand control, and methods of sand exclusion with consolidation techniques which will help the production engineer to maintain the well in a good manner so that excess production of sand does not take place.

In order to clean the well, usually nitrogen is widely used in the oil industry, purging and pressure testing the gas plant is also done which brings the well into production. With the help of nitrogen, pressure in the reservoirs is maintained which has been depleted of oxygen or has experienced reduced or low pressure.

The era of getting new oil is almost over but it is necessary to make use of efficient tools/methods/technologies to increase the production day by day without effecting the well physical and chemical behaviour. The importance of control measures is increasing day by day so that reduction of pumping of acids will take place which will help the wells strength and reduce expense.

This book presents procedures taken in the oil and gas industry in identifying the well production problems with solutions. Additional support to readers is provided with the help of various case studies.

List of Figures

Figure 1.1	Cryogenic temperature scale.	4
Figure 1.2	Nitrogen pumper.	5
Figure 1.3	Oil field nitrogen unit.	6
Figure 1.4	N_2 down annulus.	13
Figure 1.5	N_2 down tubing.	14
Figure 2.1	Relative Permeability Curves For Water And Oil.	33
Figure 2.2	Example Water Production History Curves.	37
Figure 2.3	(a) Water Coning (b) Water cresting.	39
Figure 2.4	FLow pattern of a waterflood through (a) homogeneous permeability and (b) a zone with a high-permeability streak.	40
Figure 2.5	Crossflow Bypassing a Block in a High Permeability Streak.	45
Figure 2.6	(a) Flow Pattern of a Waterflood Through a Zone With a High- Permeability Streak. (b) High-Permeability Streak Treated Before Watering Out: Injector Treated. (c) High-Permeability Streak Treated Before Watering Out: Producer Treated.	47
Figure 2.7	Example Well Pattern: Choosing Injectors.	48
Figure 2.8	Sliding Sleeves Added To Completion To Allow Future Zonal Isolation.	49
Figure 2.9	Multilateral Well Completed To Accommodate Reinjection Rather Than Lifting of Produced Water.	50
Figure 2.10	Treatment Effect on Example Well A-01.	53
Figure 2.11	How a Relative Permeability Modifier Might Adjust Relative Permeability Curves.	55
Figure 2.12	Predicted Heat-Up Profile of a Treated Well: Treatment Injection Time = 12 Hr, BHT = 91°C, Treatment Radius = 7 M.	65
Figure 3.1	Surface Valve, Eroded by Sand Production.	70

Figure 3.2	All-Welded, Pipe-Base Wire-Wrapped Oilwell Screen.	84
Figure 3.3	Casing-External Prepacked Gravel-Pack Screen.	85
Figure 3.4	A Typical Sieve Analysis Plot Showing the Value of the D_{50f} Point (in.), Found by Extending a Perpendicular Line Down from the D_{50f} Point on the Curve.	101
Figure 3.5	Effect of D50p/D50f Ratios on Sand Control and Pack Permeability, after Saucier (1974).	102
Figure 3.6	Relationship of Pack to Formation Grains at $D_{50p}/D_{50f} = 5$.	103
Figure 3.7	Cut-Away of a High Quality All-Welded Pipe-Base Wire-Wrapped Screen.	104
Figure 3.8	Detail of Well Screen Wires Relative to the Vertical Rods.	107
Figure 3.9	Well Diagram Showing the Position of Sand Packed in the Annulus.	108
Figure 3.10	Well Diagram Showing the Position of Sand Packed in the Rathole.	109
Figure 3.11	Well Diagram Showing the Position of Sand Packed in the Perforation Tunnels.	110
Figure 3.12	Sand Placement for the Entire Gravel Pack.	111
Figure 3.13	Use of Computer Software to Generate an Optimum Gravel-Pack Design.	112
Figure 3.14	Sand Grains Locked Together by In-Situ Resin Consolidation, Leaving Pore Spaces Open to Flow.	113
Figure 3.15	Application of In-Situ Resin Consolidation Steps for Internally Hardened System.	115
Figure 3.16	Application of Resin-Coated Sand Slurry.	119
Figure 3.17	After the sand grains are locked together, the sand in the casing is drilled out, leaving the wellbore clear.	120

List of Tables

Table 1.1	N_2 Pumper Population in India	2
Table 2.1	Systems Used to Affect Permeability and to Treat Specific Water-Production Problems	57
Table 2.2	Comparison of Treatments	62
Table 3.1	Merits and Limitations of Sand Consolidation Methods	81
Table 3.2	Common Polymers Used in Gravel-Packing	87
Table 3.3	Gelled Fluids and Their Characteristics	88
Table 3.4	Standard Series of 12 Sieves for a Sample Analysis	100
Table 3.5	Recommendations for Pack-Sands and Screen Devices	103
Table 3.6	Recommended Screen Diameters for Adequate Gravel-Pack Annulus	106
Table 3.7	Job Calculation Worksheet	107
Table 3.8	Carrier Fluid Requirement & Slurry Volume to Place 10 ft^3 Sand	112

List of Abbreviations

CGA	Compressed gas association
ESP	Electrical submersible pump
F	Fahrenheit
HCL	Hydrochloric acid
K	Kelvin
Kpa	Kilo pascal
N_2	Nitrogen
PI	Productivity index
RPM	Rotation per minute
S	Skin factor
WOC	Water oil contact
WSS	Well stimulation service

1

Nitrogen Application

1.1 Introduction

Nitrogen, or N_2, is the main part of the air. It constitutes 75% of air by weight or 79% by volume. Almost all the rest is oxygen.

N_2 is one of over a hundred known chemical elements. An element is a chemically pure substance containing molecules of only one type. Some elements are found in nature as gases, others occur naturally as liquids or solids. N_2 is found in nature only as a gas. Hydrogen, oxygen and helium are other commonly known elements, which are found naturally as gases. When it was discovered in 1772, N_2 was one of the first gaseous elements to be isolated. Now it is known to be the most common gas on earth and generally inert (meaning that it does not easily combine chemically with other elements).

When an inventive oilman named Duke Bloom first introduced gaseous N_2 to the oil industry in 1952, few people realized that so many uses would develop for this common gas.

The fertilizer industry is the largest consumer. Larger amounts of N_2 gas are used by the electronic industry, which uses the gas as a blanketing medium during the production of components such as transistors and diodes. Large quantities of N_2 are used in annealing stainless steel and other steel mill products. N_2 is used as a refrigerant, both for the immersion freezing of foods and during the transportation of food products. The semen used in artificial insemination is stored and transported in a liquid N_2 medium.

Liquid N_2 is used in missile work for purging components and insulating space chambers. Liquid N_2 is also used in the field of research for the development of superconductors for the flow of electricity.

Liquid N_2 is also used extensively in the oil industry. Oil field uses of N_2 are for purging and pressure testing gas plants, cleaning out wells, bringing wells into production and for foaming of acid and Frac liquids, etc.

Table 1.1 N_2 Pumper Population in India

Region	Type	Make	No.
Western Region	Skid Mounted	In-house	One
	Truck mounted	S&S	Five
	Truck mounted	Nowsco	One
	Truck mounted	Hydra	Two
Southern Region	Truck mounted	S&S	Two
	Trailer mounted	Fracmaster	One
	Skid Mounted	S&S	Two
	Truck mounted	S&S	Three
Eastern Region	Truck mounted	Nowsco	One
	Trailer mounted	Nowsco	One
	Skid Mounted	Nowsco	One
Mumbai Region	Skid Mounted	HPT	One
	Skid Mounted	Hydra	One
Total			Twenty Two

1.2 History of N_2

In 1772 gaseous N_2 was discovered by Daniel Rutherford, a Scottish physicist, and independently by Swedish Chemist Karl Wilhelm Scheele. Scientists were subsequently unable to liquefy this new gas for over 100 years.

In 1883, N_2 was liquefied by Wroblewski and Olszewski by the application of combined compression and refrigeration to temperatures below −223 °F.

N_2 was introduced to the oil industry by Duke Bloom of Bloom Aircushion Corporation of Bakers field, California. A 'Tube Transport' which was a truck carrying tyres of high pressure N_2 cylinders were used to supply gaseous N_2 to cushion a drill stem test in the Arvin field, California. A new industry was born.

In 1957, a liquid N_2 pump and vaporizer system was developed in a Cambridge Massachusetts laboratory to operate at 10,000 psi. Design considerations for larger pumps were proven technically feasible.

John F (Spi) Langston of Denton Spencer Company, introduced the first use of N_2 to the Canadian oil field and industry in 1959.

In the same year, Paul Duron, an engineer from California built first high volume liquid N_2 pump. This was subsequently manufactured by Airco Cryogenics. Although bigger pumps are now available the basic design remains essentially unchanged today.

Bloom introduced the first liquid N_2 unit in 1960, using an insulated liquid tank, a cryogenic pump and a vaporizer. This same type of equipment has been in use since 1946 for liquid oxygen ('Cascade System' for hospital bottle refill) but had not been used for N_2.

1.2.1 N_2 Properties

Liquid N_2 is lighter than water. One liter of water weighs 1.0 kg/whereas one liter of liquid N_2 weighs only 0.809 kg. Gaseous N_2 is lighter than air. At $20\,°C$, 1 m^3 of air weighs 1.205 kg, whereas 1.0 m^3 of nN_2 weighs 1.165 kg. To give a specific gravity (to air) of 0.967.

Other Properties are:-

- Colorless, Odorless, tasteless, inert, non-corrosive and non- toxic.
- Neither supports combustion, nor respiration.
- Chemical symbol: N_2
- Molecular weight: 28.0134
- Boiling point: $-195\,°C$
- Melting point: $-210\,°C$
- 1 m^3 of liquid N_2: 694.43 m^3 of gaseous N_2 (at $15\,°C$ and at 1 atm.)

For use in oil field work, N_2 is liquefied and transported in vacuum insulated tanks. There are several reasons for transporting N_2 as a liquid and they are as follows:

1. Liquid N_2 is readily available from a number of suppliers (viz. Indian Oxygen Ltd. & Bombay Oxygen at Bombay) as it is a by-product of the manufacturing process of industrial gases and is obtained from the air by separation process used to obtain liquid oxygen.
2. 1 m^3 of liquid N_2 when converted to gas will give approx. 695 m^3 of gas. This means that a relatively small amount of liquid N_2 will convert to a much larger volume of N_2 gas making it much more economical to transport.
3. In its liquid form, N_2 is pumped through positive displacement pumps up to 10,000 psi, whereas for the same quantity and pressure, we need a huge compressor for pumping gaseous N_2.

1.3 Cryogenics

1.3.1 Introduction

The word cryogenics means, literally, the production of icy cold; however, the term is used today as a synonym for low temperatures. The point on the temperature scale at which refrigeration in the ordinary sense of the

4 Nitrogen Application

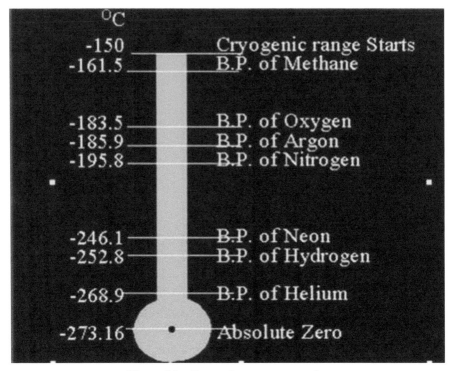

Figure 1.1 Cryogenic temperature scale.

term ends and cryogenics begins is not sharply defined. The workers at the National Bureau of Standards at Boulder, Colorado, have chosen to consider the field of cryogenics as those involving temperatures below −150°C (123K or 240°R). this is a logical dividing line, because the normal boiling points of the so-called permanent gases, such as helium, hydrogen, neon, N_2, oxygen and air, lie below −150°C, while the Freon refrigerants, hydrogen sulfide, ammonia and other conventional refrigerants all boil at temperatures above −150°C.

In the field of cryogenic engineering, one is concerned with developing and improving low-temperature techniques, processes, and equipment. As contrasted to low-temperature physics, cryogenic engineering primarily involves the practical utilization of low-temperature phenomena, rather than basic research, although the dividing line between the two fields is not always clear-cut. The engineer should be familiar with physical phenomena in order to know how to utilize them effectively; the physicist should be familiar with engineering principles in order to design experiments and apparatus.

A system may be defined as a collection of components united by definite interactions or interdependencies to perform a definite function. In general, we shall use the term cryogenic system to refer to an interacting group of components involving low temperatures. Air liquefaction plants, helium refrigerators, and storage vessels with the associated controls are some examples of cryogenic systems.

1.4 Basic Equipment

An oil field unit may be mounted on either a semi-trailer or on a single chassis truck or on the skid. Besides the truck, it comprises of three basic components:-
1. An insulated vessel that carriers liquid N_2.
2. A cryogenic pumping system
3. A vaporizer unit to convert the liquid N_2 into a gaseous form.

Figure 1.2 Nitrogen pumper.

6 *Nitrogen Application*

The balance of the equipment of the N_2 unit consists of piping, instrumentation and controls.

Figure 1.3 Oil field nitrogen unit.

The insulated vessel in construction can be regarded as being familiar with a vaccum flask. They are double walled construction, the inner tank being stainless steel and the outer mild steel. The annular space between two tanks contains insulating materials under a vacuum to reduce heat transfer.

The high-pressure liquid N_2 pumps are of a positive displacement type and are usually in a triplex configuration. The PD pumps raise LN2 pressure to that required to carry out the particular job undertaken. The liquid N_2 thus pumped is vaporized to a gas by the addition of heat either by a direct fired or a flameless vaporizer.

1.4.1 Storage Tank

Liquid N_2 is stored in a special pressure vessel at $-195\,°C$ ($-320\,°F$). The vessel is actually no more than a large twin-shelled 'thermos bottle'. The inner tank is a stainless steel pressure vessel that holds the liquid N_2. While the outer steel shell provides the insulating space on which a near vacuum is drawn. The vacuum that is applied is usually 5.0 millitorr. One torr is 1.33×10^{-3} bars. Approximately 8" of special 'perlite' insulation and polished inner surfaces further help reduce radiant heat transfer. Liquid N_2 in storage does lose a small percentage of product continuously to vaporization but can be stored for an extended number of weeks with minimal product loss.

1.4.2 Pumping System

Liquid N_2 from the tank flows to a 'boost pump' through stainless steel pipes. This boost pump is a hydraulic centrifugal pump with cryogenic liquid handling capabilities. The boost pump raises the pressure of the liquid up to 827 Kpa (120 psi). The N_2 at 827 Kpa (120 psi) is then fed to a high-pressure cryogenic pump. This pump is a positive displacement 'Triplex' pump which raises the pressure to the desired level. From the high-pressure pump, the liquid N_2 is forced through a series of stainless steel coils which are heated by hot air from a diesel burner. The liquid N_2 in these coils absorbs heat and is gasified. This duct which flows down the line for its various service applications.

1.4.3 Vaporizer System

On the type of vaporizer system, N_2 units are categorized as;
- Diesel fired N_2 pumping unit
- Non-fired N_2 pumping unit.

In the former type of unit, liquid N_2 from high pressure pump is forced through a series of stainless steel coiled tubes which are heated by hot air from a diesel burner. The liquid N_2 in these coils absorbs heat and is gasified.

The non-fired N_2 pumping unit works on the waste heat recovery principle. This vaporizer system uses the engine coolant (glycol/water) to recover heat from the engine, transmission, exhaust and hydraulic system. Heat may artificially be enhanced by drawing power from a diesel engine by means of 'Allison transmission retarder'. This is the heart of the non-fired N_2 vaporization system.

The engine coolant temperature is about 76–82 °C (170–180 °F) as it leaves the engine system. Heat exchange between this coolant and liquid N_2 from high-pressure triplex pump takes place in a helical coiled tube heat exchanger.

Low temperature coolant is then circulated through a transmission coil to a coolant heat exchanger and exhaust gases to a coolant heat exchanger to reach about 60 °C (160 °F) before it re-enters the diesel engine.

The temperature of the gasified N_2 through N_2 to coolant heat exchanger may further be enhanced, as it passes through stainless steel coiled tubes, by diverting exhaust gases from the engine over these tubes if required, by means of a diverter valve. This gaseous N_2 from 10 to 65 °C (50–130 °F) is the final product that flows down the line for its various applications.

1.5 Safety

1.5.1 General Information

Strict compliance with proper safety and handling practices is necessary when using a Horizontal Liquid Transport Vessel.

N_2 and argon vapors in the air may dilute the concentration of oxygen necessary to support or sustain life. Exposure to such an oxygen-deficient atmosphere can lead to unconsciousness and serious injury, including death.

Before removing any parts or loosening fittings, empty a cryogenic container of liquid contents and release any vapor pressure in a safe manner. External valves and fittings can become extremely cold and can cause painful burns to personnel unless properly protected. Personnel must wear protective gloves and eye protection whenever removing parts or loosening fittings. Failure to do so may result in personal injury because of the extreme cold and pressure in the tank.

Accidental contact of liquid gases with skin or eyes can cause a freezing injury similar to a burn. Handle liquid so that it will not splash or spill. Protect eyes and cover skin where the possibility of contact with liquid, cold pipes, and cold equipment, or cold gas exists. Wear safety goggles or a face shield if liquid ejection or splashing may occur or cold gas may issue forcefully from equipment. Clean, insulated gloves that can be removed easily and long sleeves are recommended for arm protection. Wear cuff-less trousers over the shoes to shed spilled liquid.

1.5.2 Safety Bulletin from CGA (Compressed Gas Association)

Cryogenic containers, stationary or portable, are occasionally subjected to assorted environmental conditions of an unforeseen nature. This safety bulletin is intended to call attention to the fact that whenever a cryogenic container is involved in an incident whereby the container or its safety devices are damaged, good safety practices must be followed. The same holds true whenever the integrity or function of a container is suspected of abnormal operation.

Good safety practices dictate that the contents of a damaged or suspect container be carefully emptied soon as possible. Never leave a damaged container with product in it for an extended time. Do not refill a damaged or suspect container unless the unit has been repaired and the problem rectified.

Incidents that require that such practices be followed include highway accidents, immersion in water, exposure to extreme heat or fire, and exposure to most adverse weather (earthquakes, tornadoes, etc.). Whenever a container is suspected of abnormal operation or has sustained actual damage, follow good safety practices.

If there is known or suspected container vacuum problem (even if an extraordinary circumstance, such as those noted above, has not occurred), do not continue to use the unit. Continued use of a cryogenic container that has a vacuum problem can lead to embrittlement and cracking. Further, the carbon steel jacket could possibly rupture if the unit is exposed to inordinate stress conditions caused by an internal liquid leak.

Before reusing a damaged container, always test, evaluate, and repair as necessary. It is highly recommended that any damaged container be returned to Hydra Rig for repair and rectification.

The remainder of this safety bulletin addresses those adverse environments that may be encountered when a cryogenic container has been severely damaged. These are oxygen-efficient atmospheres, oxygen-enriched atmospheres, and exposure to inert gases.

1.5.3 Oxygen-deficient Atmospheres

The normal oxygen content of air is approximately 21%. Depletion of oxygen content in the air, either by combustion or by displacement with inert gas, is a potential hazard, and users should exercise suitable precautions.

One aspect of this possible hazard is the response of humans when exposed to an atmosphere containing only 8–12% oxygen. In this environment, unconsciousness can be immediate with virtually no warning.

When the oxygen content of air is reduced to about 15 or 16%, the flame of ordinary combustible materials, including those commonly used as fuel for heat or light, may be extinguished. Somewhat below this concentration, an individual breathing the air is mentally incapable of diagnosing the situation because the onset of symptoms, such as sleepiness, fatigue, lassitude, loss of coordination, errors in judgment and confusion, can be masked by a state of 'euphoria', leaving the victim with a false sense of security and well-being.

Human exposure to an atmosphere containing 12% or less oxygen leads to rapid unconsciousness. Unconsciousness can occur so rapidly that the user is rendered essentially helpless. This can occur if the condition is reached by an immediate change of environment or through the gradual depletion of oxygen.

N_2 and argon (inert gases) are simple asphyxiates. Neither gas will support or sustain life and can produce immediate hazardous conditions through the displacement of oxygen. Under high pressure, these gases may produce narcosis although an adequate oxygen supply, sufficient for life, is present.

N_2 and argon vapors in air dilute the concentration of oxygen necessary to support or sustain life. Inhalation of high concentrations of these gases can cause anoxia, resulting in dizziness, nausea, vomiting, or unconsciousness and possibly death. Prohibit individuals from entering areas where the oxygen content is less than 19% unless equipped with a self-contained breathing apparatus. Unconsciousness and death can occur with virtually no warning if the oxygen concentration is less than approximately 8%. Contact with cold N_2 or argon gas or liquid can cause cryogenic (extremely low temperature) burns and freeze body tissue.

Immediately move anyone suffering from lack of oxygen to areas with normal atmospheres. self contained breathing apparatus may be required to prevent asphyxiation of rescue workers. Assist with respiration and supplemental if the victim is not breathing. If cryogenic liquid or cold boil-off gas contacts a worker's skin or eyes, promptly flooded or soak affected tissues with tepid water (105–115 °F; 41–46 °C). Do not use hot water. Have a physician promptly exam cryogenic burns resulting in blistering or deeper tissue freezing.

1.5.4 Safety for Handling and Exposure

Any contact of liquid or cold gaseous N_2 with any body part can induce freezing similar to frostbite. When using N_2, it is essential to use proper body protection and dress when handling or in and around its use.

Avoid splashing or spillage when handling liquid N_2 (LN_2). Wear safety goggles and clothing that covers as much of the body as possible. Avoid clothing, such as cuffs in pants, and shirtsleeves or pants tucked inside boots that could 'catch' liquid N_2. Gloves that are insulated, easily removed and cleaned are preferred to ordinary gloves.

Although bodily protection is important, a well-ventilated work environment is even more essential. N_2 is odorless, colorless and tasteless. The human senses cannot detect it within the air. Without proper ventilation, N_2 can expand, displacing enough oxygen in the working atmosphere to cause dizziness, unconsciousness or even

death when inhaled. Store liquid N_2 (LN_2) containers in well ventilated facilities or outdoors.

To summarize safety using gaseous or liquid N_2, always remember the following:

1. Keeps work area well ventilated.
2. Dress appropriately.

1.6 N_2 Service Applications

A list of the uses of N_2 in the oil field in being given here. However, only some of important uses are described here in details

1. Displacement
2. Drill stem testing
3. Hydro perforation
4. Nitrified fluids – Acidization
5. Atomized acid
6. Formed acid

7. Aerating conventional fluids
8. Pressure testing
9. Pipeline purging
10. L V O cementing technique using N_2
11. Tubular inhibition treatments
12. In mud systems to reduce lost circulation
13. Freeing differentially stuck drill pipe
14. Relieve water blockage
15. N_2 test for communications
16. N_2 to set hydraulic packers
17. Casing insulation
18. Treatment of injection wells
19. Scale removal with foamed acid
20. Freeze plug
21. N_2 to control annular pressure
22. N_2 with air drilling
23. Use of foam as a drilling and workover fluid
24. Foam clean out
25. Blow out stimulations
26. Development of an off shore gas condensate reservoirs
27. Fire control
28. Water control technique by N_2 injection

1.6.1 Displacement

The use of N_2 as a medium for displacing well fluids has proven itself as a safe and economical method for completions, tests, recovering inhibitors and spent stimulation fluids.

The N_2 line is connected to the annulus side, usually through the casing valve. N_2 is then pumped down the annulus forcing the fluid up and out the tubings. The depth to which the fluid is displaced depends upon the pressure method of displacing fluids from the well.

Circulation of N_2 down tubing with returns from annulus is sometimes preferred because less N_2 volume is required to circulate complete well to N_2. With N_2 lifting from below the well, the maximum pressure is reached when N_2 is at the bottom of the tubing.

For zone evaluation, the fluid is normally displaced to a depth close to the bottom of the tubing. After displacing, the tubing is lowered into the packer or the packer is reset. The N_2 then can be bled off at a controlled rate and well

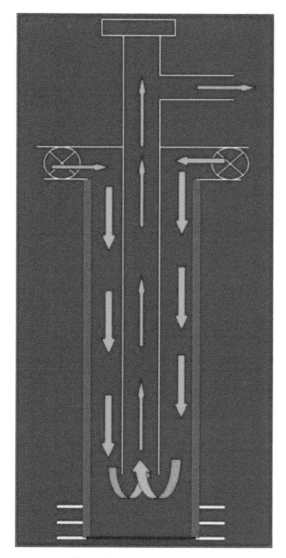

Figure 1.4 N_2 down annulus.

evaluated. Caution should be used as there will be considerable N_2 pressure at the return line behind the fluid.

Displacing to perforate is basically the same procedure. The well fluid is displaced to a depth that is adequate to ensure the hydrostatic weight of the fluid remaining in the tubing is less than the anticipated reservoir pressure.

14 *Nitrogen Application*

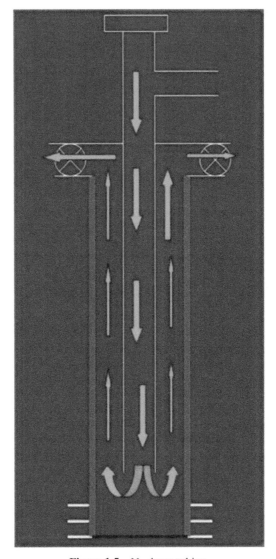

Figure 1.5 N_2 down tubing.

After a well is perforated, the N_2 pressure can be bled off at a controlled and zone evaluated.

Displacing both tubing and casing is calculated from the tables, nicknamed 'Zip Tables', showing N_2 gas volumes over a large range of pressure in a well.

Another useful method of displacement is bull-head displacement method in which the fluid is forced into the formation rather than being circulated out of the annulus. The most common reason for displacing in this manner is to save the cost of a rig to unseat the packer to enable circulation. Also, stimulation fluids and other chemicals are displaced to the formation using this method.

On a typical job, the N_2 discharges line is connected to the wing valve or top of the X-mas Tree. Using a known or an assumed bottom hole pressure, the surface pressure and N_2 requirements are calculated. During the job, the depth at which you have displaced can be calculated by comparing the surface to the amount of N_2 pumped. On most jobs, we will get a pressure drop when N_2 reaches the formation due to the fact that it is easier to pump gas into a formation than it is to pump liquid into it.

Sample problems are given below: Example 1

> What volume of N_2 would be required to completely displace a well fluid with 3% KCl water?
>
> Depth – 2200 m Tubing – 73 mm (2.7/8")
>
> Casing – 177.8 mm (7")
>
> N_2 is to be injected down annulus with return from tubing.

The addition of 3% KCl to water does not change its density much from that of water (density of 3% KCl is 1017 kg/m^3, compared to 1000 kg/m^3 for water), so, return to page 'Fluid water' in the ZIP Tables

Under the column 73–177 mm ANN read the answer across from 2200 m depth volume – 5984 m^3 N_2 gas.

1.6.2 Nitrified Fluids-Acidisation

One of the time and money-consuming factors concerning acid stimulation is well clean-up after the job. For this reason, one of the main advantage of using N_2 in conjunction with acid stimulation is the recovery of the fluids injected into the formation without the extended and hazardous conditions involved by swabbing.

N_2 has low solubility in fluids. Therefore, most of the N_2 commingled with the acid is in the state of compressed gas bubbles. These compressed bubbles serve as sources of energy within the fluids injected into the formation. This energy may then be expended as an additional force when the pressure is relieved as the well is opened up. This additional energy, supplementing the existing formation energy, will greatly aid in the removal of the fluids to the wellbore and to the surface. This added velocity will also greatly increase the removal of insoluble precipitates and formation fines along with acid is the increase in volume due to the addition of the N_2. The radius of penetration of the fluids away from the wellbore is relative to the volume injected at bottom hole conditions.

In order to obtain the maximum benefit from using N_2 commingled with fluids injected into a well, it is important that each job be individually designed. By properly considering the various well variables and combine this information with the data contained in the Technical Manual, the proper ratio of gas to liquid for a particular well may be quickly determined.

1.6.3 Atomized Atom

Atomized acid is a fine mist of acid droplets in a continuous gas phase (N_2). Atomization is accomplished by injecting a liquid into a high-velocity stream of N_2 gas by means of a specially designed atomized chamber. The gas-liquid ratio is adjusted so that at bottom hole conditions, the N_2 gas is in a continuous phase of two-phase treating fluids.

In treating oil and gas formations with liquids, such as during acidizing, a problem of liquid blockage can occur in the vicinity of the wellbore. This condition occurs when the injected liquid wets a portion of the reservoir rock matrix and the resulting capillary forces holding the trapped liquid are greater than the flow shear forces holding the trapped liquid which attempt to dislodge it. Under these conditions the trapped liquid becomes immobile and serves to block the flow of the normal reservoir fluids, thereby restricting flow into the wellbore.

Also in many formations, a liquid blockage already exists in the vicinity of the wellbore. The source of this liquid could be from original drilling mud filtrate, accumulation of produced formation fluids, or from previous liquid treatments.

In order to avoid the creation of this problem, it would be preferable to inject the required treating liquid into the formation as a fine mist of liquid droplets in a continuous inert gas phase such as N_2. This is accomplished with atomized acid techniques. The benefits derived from atomization are as follows:

1. Avoidance the formation of a water block by injecting a continuous gas phase rather than a liquid phase.

2. The high velocity of the expanding N_2 in the formation during flow-back aids in cleaning up the precipitates and formation fines in the vicinity of the wellbore.
3. The high mobility of the N_2 permits the entry of acid into the smaller rock interstices that would be blocked by capillary forces during conventional treatments.
4. Increased areal coverage for the same acid volume due to the expanded volume of the atomized system.
5. Better penetration into, and subsequent removal of existing liquid blocks due to the large mobility ratio of the gas relative to the liquid.

An analytical development and analysis of field problems indicate the feasibility of injecting into formations a liquid phase dispersed in a continuous N_2 gas phase. These tests indicate that not only the process is feasible but the results obtained using atomized acid are generally improved over conventional acid treatment.

The desired gas to liquid ratio is dependent on several factors. The single most important factor is the desired accomplishment of the atomized acid treatment. If the problem is immediate wellbore blockage, a ratio of 1:1 or 2:1 should be sufficient. On the other hand, if it is desired to obtain a larger areal coverage of the formation with a given volume of acid, a larger expansion ratio such as 4:1 to 6:1 should be used. The amount of N_2 always depends upon the bottom hole pressure in calculating expansion ratios.

One of the biggest problems in designing an acid treatment especially in low permeability or low-pressure reservoir is a recovery of spent acid. The longer the acid is allowed to remain in a well, the greater are the chances of formation damage and tubular damage. In many cases, the spent acids will have to be swabbed before the well can unload fluid continuously.

1.6.4 Foamed Acid

In foamed acids, 15%, 20% or 28% HCl is used as the water base, continuous phase. The following discussion describes several possibilities as to why foamed acid has been so successful.

1.6.4.1 N_2 Retention

Aerated acid has been used for many years. An aerated acid usually contains 500–1000 SCF N_2 per barrel of acid. The N_2 is added to lighten the fluid head, allowing many wells to flow back naturally. However, under certain circumstances, N_2 may become separated from the acid and 'lost' in the formation. These wells need usually swabbing to clean up the spent acid. If a surfactant is added to the acid, foam can be generated. The N_2 will then be dispersed in the acid as a discontinuous phase or microscopic bubbles. The gas is, in effect trapped within the continuous phase, which is acid.

Because the acid is the continuous phase, it should react with the formation as the foam is being pumped. With time, the foam should change from a 'live acid' foam to 'spent acid' foam. But at all times, the gas in water emulsion should remain in-tact, therefore at the end of treatment, the necessary energy required to lift the spent acid is available.

1.6.4.2 Diverting

Even in areas where spent acid recovery is not a problem, foamed acid can offer several advantages over conventional treatments. Foam has been recognized as a fluid that can be used to divert acid from one set of perforations to another. In massive formations, an acid treatment will tend to enter the most permeable or the lowest pressure section, leaving some or most of the interval un-stimulated.

The viscosity of foam can be several hundred centipoises depending on the shear rate. Therefore, by running a pad of foam between the two stages of acid, the increased viscosity of the foam should provide a sufficient wellbore pressure increase to effectively divert the second to a new interval.

If all the acid is foamed prior to pumping it into the well, then the foamed acid should have 'inherent' diverting properties. When the foamed acid enters a set of perforations and moves into the formation, the injection pressure should begin to increase. Finally, the pressure will reach a point where the foamed acid will break down another set of perforations. The process should continue until each zone has been acidized.

1.6.4.3 Production of Fines

Another potential benefit of foamed acid, which applies to all formations, concerns the solids carrying capacity of foam. Probably the two most widely used applications of foam are:

(1) Wellbore clean out and (2) foam fracturing. Both of these processes take advantage of the excellent suspension of solids, which is another property of foam.

During acid treatment, there is normally a large amount of fines that are released. Many of these fines will be 'picked up' by the foam and returned to the surface. This fact alone may mean the difference between success and failure for many treatments.

1.6.4.4 Foamed Acid Guidelines
1. To obtain a foamed acid, use 10 gallons of surfactant per 1000 gallons of acid.
2. Never use less than 5 gallons of surfactant per 1000 gallons of acid.
3. For matrix acid jobs, a foam quality of 70% will provide a minimum foam viscosity.
4. For acid fracturing treatments, a foam quality of 80–90 % will provide a maximum foam viscosity
5. It is impossible to generate foam in the laboratory at low rates without the use of a foam generator. Therefore, for low rates, foamed acid treatments (when the flow down the tubing is laminar), a foam generator or atomizer in the injection line may be beneficial.

1.6.5 Aerating Conventional Fluids

Hydraulic fracturing of the reservoir formation has been performed for years as a means of increasing production. The formation is fractured by means of a fluid being forced into it hydraulically. The created fractures are new channels for the flow of formation fluids and/or gases. Normally, some type of proppant such as sand, glass aluminum beads, or walnut shells, is mixed with the fluid to keep the fractures propped open when the hydraulic pressure is dropped. The formation tries to heal or return to its original shape when pumping is stopped but is prevented from doing so by the addition of the proppant. It is common practice to fracture a formation using several thousand pounds or proppant.

Often, a very serious problem results from fracturing a formation is that the injected fluids cannot flow back to the surface. This is a common problem with low-bottom hole pressure reservoirs. A very effective method of averting this problem is to aerate the fracturing fluid, while it is being injected with N_2.

The N_2 commingled with the fracturing fluid is in the state of compressed gas bubbles. These compressed bubbles serve as a source of energy within the fluids injected into the formation. This energy may then be expended as an additional force when the pressure is relieved as the well is opened up. This additional energy, supplementing the existing formation energy, will greatly aid in the removal of the fluids towards the wellbore and to the surface.

The increased velocity of the fluids returning towards the wellbore, due to expanding N_2, will aid in the removal of crushed sand and formation fines.

Another benefit of aerating the fracturing fluid with N_2 is the built-in flow capability in the advent of a fluid pump failure while pumping sand. The well may be opened up and the fluid and sand flowed to the surface before the sand has time to fall out and bridge over. After the wellbore is cleared and necessary repairs are made, the fracturing job can be resumed.

An often overlooked benefit of using N_2 with fracturing jobs is that of fluid loss control. Data show that when N_2 is commingled with fluids at certain ratios, the fluid loss or leak-off is significantly reduced when the mixture is injected into a formation. Fluid loss or leak-off tends to block small openings in the formation and hinders flow back. Also, if enough fluid leaks off, the flow along the fracture could be reduced enough to cause a screen out.

The calculations for aerated fracturing fluids are the same as with aerated acid. The proper fluid weight curve is used if different than acid weight. The proper N_2 to fluid ratio must be determined.

1.6.6 Pipeline Purging

Inert gas that is N_2, is used for purging large vessels and pipelines. The use of N_2 eliminates internal explosion hazard during start up facility repairs. In May 1987, slug catcher phase-II of Hazira Gas processing Plant was successfully purged by WSS team with our own N_2 units.

To determine N_2 volume required for pipeline testing, it is necessary to know only three variables.

1. Pressure required in Mpa
2. Temperature in °C
3. Volume of system in m^3 (Liquid volume)

Liquid volume of pipeline = m^3 liquid/1000 m
= $0.0007854 \times $ (I.D. in m)^2

Example problem

How much N_2 in m^3 is required to pressurize 3 miles 6" I D pipe to 19 Mpa. Assume ground temperature at 10°C.

1.6.7 Use of Foam as a Drilling and Workover Fluid

Foam permits circulation in wells where water or heavier liquids would be lost to the formation. With lost circulation, the job's objective is defeated and all liquid lost must be swabbed back before production is restored.

Foam, formed by combining water, surfactant and gas has a very low density which allows circulation at very low bottom hole pressures. However, the foam maintains very good carrying capabilities.

Some rigs operating today are specially equipped to drill with foam. These rigs usually have specially designed pumps and air compressors to generate and circulate foam. However, any conventional rig is capable of using foam as the drilling or workover fluid. The single most important addition required for a conventional rig to use foam is that of a gas supply. Ideally, that gas should be N_2.

The use of N_2 gas in generating stable foam has many advantages over the use of air or natural gas:

6. N_2 is an inert gas, thus eliminating the possibility of a down hole explosion
7. Since N_2 is inert, there is no danger of fire on the surface as would be present if using natural gas.
8. Far more flexibility with rates and pressures is available when using N_2
9. Gas temperature can be adjusted to the desired circulating temperature.

Usually, the use of foam as drilling fluid is confined to one particular zones, with the rest of the well being drilled with conventional mud. Normally the drilling time to drill through a zone is relatively low, therefore, the N_2 requirements would be small. The cost of the N_2 service might be less than the cost to rent an air compressor or to lay a gas line to a gas supply. Of course, the added safety in using N_2 is difficult to price.

The normal procedure used to rig up to generate foam for drilling is somewhat different than of aerating mud. The N_2 discharge line is tied into an atomizer along with the fluid pump's discharge line. The discharge from the atomizer is connected to the stand pipe. As the fluid pump discharges the water and surfactant mixture, the N_2 is expanded into the fluid stream through the atomizer, creating foam. The foam is circulated and returned to a catch tank where the foam is allowed to break. If a means of separating solids from the fluid is provided the water and surfactant can be re-circulated as long as no hydrocarbons are present in the system.

The N_2 to fluid ratio is dependent upon operating temperature and pressure. There is a margin of choice in selecting foam density. However, the foam should contain between 65% and 90% gas by volume in order to remain stable. The choice of foam quality is largely dependent on reservoir pressures and the desired hydrostatic weight.

1.7 Foam Clean Out

1.7.1 Introduction

Hydraulically induced fractures have been used to stimulate oil and gas wells for the past many years. Hassebroek and waters summarized the advancements in fracturing technology through the first 15 years. During this period, great strides were made in the understanding, engineering and mechanical aspects of hydraulic fracturing. In the last few years, interest in hydraulic fracturing has gained a renewed momentum. The decline of domestic reserves turned the petroleum industry's attention towards the recovery of hydrocarbons from tight reservoirs. In these low permeability formations, hydraulic fracturing treatments are routinely performed upon initial completion.

Successful stimulation depends upon creating a fracture, which can be propped for the desired length, using a fluid that does not substantially reduce the formation permeability next to the fracture. The selection of the fluid, therefore, is usually the key to designing a successful fracture treatment. One of the most recent innovations in fracturing technology is the use of foam as a fracturing fluid. It was first accepted by the oil industry as fracturing fluid in 1974.

Foam, which is a mixture of gaseous N_2, water and surfactant, has been used for many years as a drilling and workover fluid. The properties of foam, such as low hydrostatic head, low water content and excellent suspension of solids, make it an ideal fluid for drilling into or working over low pressure, water sensitive reservoirs. These same properties led to the development of foam as a fracturing fluid.

The viscosity of foam was investigated by Mitchell. His experimental findings confirmed the existing theories that foam flow could be predicted using the single-phase flow theory. Blauer et al. extended Mitchell's work by investigating foam flow in oil field size tubular. The results of their work have been successfully applied to the design of hundreds of foam fracturing treatments during the past years. The properties of foam as a fracturing fluid have been well documented.

A computer program calculates the behaviour of the foam in the tubulars and resulting fracture dimensions.

Foam is like an emulsion, or gas–liquid dispersion, with gas as the internal phase and liquid as the external phase.

The foam quality is the ratio of gas volume to foam volume (volumetric gas content) at a given pressure and temperature

$$\text{Foam quality }(T, P) = \frac{\text{Volume}_{\text{gas}}}{\text{Volume}_{\text{foam}}} = \frac{\text{Volume}_{\text{gas}}}{\text{Volume}_{\text{foam}} + \text{Volume}_{\text{liquid.}}}$$

In the range of approximately 0%–52% quality, the gas bubbles in the foam are spherical and do not contact each other. Foam in this range has rheology similar to the liquid phase. In the approximate quality range of 52%–96% the gas bubbles in the foam interfere with one another and deform during flow causing the foam to increase in viscosity and yield point. In this particular range, foam behaves like a Bingham plastic fluid. A bingham plastic fluid has a yield point that must be exceeded before it flows. We see some physical evidence of yield point behaviour in high-quality foams. They resist being poured from a container and stack in rigid masses when pumped on the ground. Once a Bingham plastic yield point is exceeded, its laminar flow pressure drop is a linear function of flow rate. Plastic viscosity is the proportionality content describing this linear flow pressure drop on flow rate. Above 96% quality, foams may degenerate into a mist. The thin liquid layer is not able to contain the larger volume of gas and thus the foam bubble ruptures.

In theory, foam between 52 and 96 quality could be used to transport proppant. The higher quality foams have higher viscosity and give greater support to proppant in a static condition. However, the higher quality foams require more horsepower to pump and in the case of fracture acidizing there may not be enough acid present to effectively dissolve rock and create adequate fracture flow capacity. So a compromise is reached between 60 and 80 quality, with 70–75 quality being most frequently used in foam fracturing and 75–80 quality being most frequently used in foam acidizing.

1.7.2 Foam Stability and Viscosity

Once the foam is formed it will remain dynamically stable by keeping the foam in motion. However, if the foam stops moving, the destabilizing process of drainage occurs. Gravity forces a separation of the free liquid which is not tightly bound at the surface of the bubbles from the rest of the foam. How rapidly the static foam drains depends upon the viscosity of the liquid phase and, to some extent, on the concentration of foaming agent. As the temperature of a foam increases, the viscosity of the liquid phase decreases, so the static foam drains quicker. Generally, the concentration of the foaming agents must also be increased to stabilize foams as temperature increases.

The static stability of foams may be increased by using a gelling agent appropriate to the liquid phase. Some gel viscosity is usually desired to aid proppant transport through the surface equipment. For low-temperature fracturing treatments, that is less than 120°F, no gel viscosity is needed for foam stability with pumping times of one hour or less. In high-temperature fracturing treatments or where long pumping times are required, a viscosity of 10 cp for the liquid phase at bottom hole temperature should give adequate foam stability.

In fracturing treatments, foaming agents may be metered with liquid additive pumps and injected into the booster pump. In cases where gel has been added to the liquid and some viscosity has developed, the foaming agents may be batch mixed into the base fluid. Also, foamer may be diluted and injected into the high-pressure fluid stream.

In foamed acid treatments the foaming agent is usually batch mixed in the liquid phase.

1.7.3 Fire Control

N_2 can be used to extinguish certain well fires. If an oil and gas well catches fire, non-combustible gas can be pumped into and around the flame and produce a non-combustible mixture by lowering the oxygen content of air to less than 6% N_2 can also put out a natural gas blaze when the concentration of N_2 in natural gas can be increased above 30%.

Extinguishing the oil well fire was successfully tried in the western region in 1982 at the time of blow out of a well in the Dabka oil field.

1.8 Water Control Technique by N_2 Injection

1.8.1 Introduction

Water could be one of the constituents in produced fluids either since the beginning or in later stages of production. The phenomenon is observed in reservoirs with water drive mechanism.

Percentage of water contained in produced fluids in called as 'Water Cut' This water cut is bound to increase with time due to reasons of its own/several reasons.

Excessive water cut exerts an additional back pressure on the formation to that would have been exerted by column of fluids produced with allowable water cut or no water at all. This is undesirable due to the following reasons:

1. Reduces oil and gas production
2. Reduces well flow capability to oil and gas
3. Necessitates early artificial lift or increased artificial lift if well is already on artificial lift
4. Increase operating cost of the well
5. Early ceasure of flow in gas wells.

This obligates water cut to be reduced to a minimum

1. Using cement slurry
2. Using grouting material

    (i) Sodium silicate slurry
    (ii) Plastics/polymer plug

3. Using water soluble polymer

    (i) Ployacrilamides
    (ii) WORCON
    (iii) AQUTROL I
    (iv) K – TROL

4. Using gaseous N_2

Among the various techniques mentioned above, all other techniques, except 'Using N_2', 'N_2', are out of the scope of our syllabus / discussion.

1.8.2 Technology

N_2 in gaseous foam enters water channels preferentially due to the difference in viscosities of oil and water and pushes water back far away into the formation. The N_2 being non-wetting phase occupies larger pore spaces and water occupies smaller pores. Thus, entrapped in the larger channels gaseous N_2 acts as a blockage to the movement of water.

The surfactant used, which is essentially water wet, further aids in reducing relative permeability to water.

Leased oil pumped for cushion not only pushes N_2 further deep into the water channels but also restabilizes oil saturation around the well bore and inter connects oil channels which were previously by- passed by water. Thus, restored permeability to oil enables the by-passed oil to flow towards the well bore.

1.8.3 Job Description

A typical job would recommend pumping a calculated amount of N_2 with surfactant. The Thumb rule for N_2 is 2 to 3 m^3 of liquid N_2 per meter of pay zone. Surfactant in N_2 does not exceed 4% of liquid N_2 volume in liters. Sometimes, N_2 pumping is preceded by crude pumping which has the water wet surfactant in quantity not exceeding 4% of oil volume in litres.

Calculated quality of leased crude is then pumped and the well is closed for establishment (generally 24 hours.). then well is flowed through choke at largest particle size (ϕ 3–6 mm).

Care is to be taken to optimize the fluid volume to be pumped based on the trial done on the field where this method is proposed.

1.8.4 Commercial Viability

Following features could be adumbrated for water shut-off by pumping N_2.

1.8.5 Quick and Easy

It eliminates cumbersome and time consuming operations like mixing and preparing a slurry of required gelling properties. Further, this or treatment time required is few hours (24–48 hours).

1.8.6 Versatility and Adaptability

This method could be applied to reservoirs having a high permeability to those with low permeability. This method can easily be coupled with other pumping methods.

1.8.7 Economical

Due to above mentioned reasons, it reduces operating costs greatly. Further, major factor in cost is N_2 which is very less. Also this method gives a dual benefit of increased oil production and reduced water production which reduces effluent treatment cost.

But it is worth noting here that this method does not offer a permanent solution to the problem. Further 'Longevity' of the job could be less compared to the offered by conventional methods. This application has been found successful, particularly for coning, fingering and cusping of water in the reservoir.

1.8.8 Freeding Differentially Stuck Drill Pipe

To free the pipe, it is necessary to decrease the hydrostatic pressure exerted on the formation allowing the formation to heave or kick toward the well bore, thus freeing the pipe. On a typical job this is done in the following manner Connect N_2 line to kill line on well head control and close blowout preventor. Displace sufficient drilling fluid down the annulus out drill pipe and as N_2 is bled off and fluid U-tubes from the drill pipe and annulus its level will exert the same hydrostatic pressure as was exerted when well kicked prior to sticking. As soon as N_2 is bled off, attempt to pick up drill pipe & if it is free, pull into a good hole, if casing is near are of sticking, pull up into casing and circulate N_2 and air bubbles out of hole. If after N_2 is bled off, a pipe does not respond to decrease in hydrostatic pressure in well bore, Operator should continue to circulate hole until N_2 is ready to be pumped, otherwise solids in the drilling fluid will settle and plug the bit.

At times areas of porosity or fractures charged with pressure may be encountered. When this occurs, it may be necessary to increase the drilling fluid weight to control the well. if this is done and the well is balanced or near balanced, drilling usually continues.

Problem: Calculate depth to displace to free pipe

Reservoir pressure	1500 psi
Reservoir depth	3000 ft
Mud weight	12 ppg
Drill pipe size	3 × 1/2 inches
Hole size	7 × 5/8 inches

Solution:

Step 1 Calculate depth so you have a 200 psi differential from formation to well bore D = reservoir pressure − 200 psi 4 mud weight × 0.052

Step 2 Subtract depth calculated from depth to formation to get depth to displace.

Calculation:

Step 1 Depth = 1500 psi − 2004 12 ppg × 0.052 = 1300 psi 4.624 = 2083'
Step 2 3000' − 2083' = 917'

This calculation does not consider the fact that the drill pipe fluid column and the annulus column will seek a common depth when the N_2 is bled off. Since the annuals volume is 4.3 times that of the drill pipe, when the two strings equalize the depth to fluid will be 744'. The pressure on the formation would therefore be 3000' – 744' × 12 ppg × 0.052 = 1408 psi. This should still be enough differential to free the pipe.

1.8.8.1 N_2 Lift

N_2 gas circulated into the production conduit to displace liquids and reduce the hydrostatic pressure created by the fluid column. N_2 lifting is a common technique used to initiate production on a well following Workover or overbalanced completion. A coiled tubing string is generally used to apply the treatment, which involves running to depth while pumping high-pressure N_2 gas. Once the kill-fluid column is unloaded and the well is capable of natural flow, the coiled tubing string is removed and the well is prepared for production.

1.8.8.2 N_2 cushion

High-pressure N_2 is typically applied to a tubing string in preparation for drill stem testing or perforating operations in which the reservoir formation is to be opened to the tubing string. The N_2 cushion allows a precise pressure differential to be applied before opening flow from the reservoir. Once flow begins, the N_2 cushion pressure can be easily and safely bled down to flow formation fluids under a high degree of control.

1.9 Case Study I

Well No.	:	S #158
Reservoir Name	:	ABC
Type of Reservoir	:	Water drive

Reservoir Details:

Sanand is one of the older fields. It is s matured field supported by a weak aquifer. The field is producing for more than 35 years and hence the water cut is high. As the viscosity of the oil from this field is very high and the reservoir pressure is depleted, most of the wells are on SRP. The weak aquifer has not been supported to enhance the pressure substantially.

The well under consideration is producing from newly developed sand and the results are not encouraging as the viscosity of the oil is too high. But it opened the way for further prospects.

Problem in Well : No influx from the reservoir into the well

Well Data
   10. Tubing size : 2 7/8″
   11. Casing size : 5 1/2″
   3. PI : 1526–1529

Job Type : Bottom Cleaning and Activation

Principle Applied:
A. Bottom Cleaning

1. Jetting action of water through jetting tool attached at the end of CT helps break the obstruction and entraining the particles in the circulating medium.
2. The thixotrophic property of the circulated gel water helps in transporting the entrained particles from wellbore to the surface.

B. Activation

1. Liquid N_2 is used for well activation purposes. It eventually is reconverted into a gaseous form which occupies a much higher volume. This increase in volume along with the release of energy during the reconversion process applies enough pressure on the fluid in the wellbore, pushing it out to the surface.

Job Execution:
A. Bottom Cleaning

1. The CT (1 ¼″) is run inside the hole and water is circulated at 1000 psi up to 1400 m
2. The CT is lowered further with gelled water circulation at 3500 psi up to 1545 m.
3. No obstruction was felt.
4. CT is pulled out of hole.

B. Activation

1. CT is again lowered with N_2 gas at 2100 psi up to 1275 m.
2. Liquid in the wellbore is knocked out from 600 to 1278 m.
3. When only N_2 is observed in the return, the CT is pulled out of hole.

1.10 Results/Remarks

After bottom cleaning, frac fluid, sand and water are observed in the return. After activation, frac fluid, water and N_2 gas is observed in the return.

1.11 Conclusion

N_2 operation for well services has been an integral component in the global Oil and Gas Industry. N_2's usage for activation and foaming purposes is widely accepted due to its inherent properties. Its non-toxic, non-inflammable and non-corrosive behaviour helps in prolonging the life of equipment and also adds to the safety factor. Being an easily available inert gas, it is the ideal choice for most operations.

N_2 is and will continue to be an essential part of the industry and a good understanding of its properties and applications will go a long way in enabling the user to exploit its full potential.

1.12 Specification of N_2 Pumpers Available with WSS

COLD END

Type/mode	1-LMPD	3-GUPD	1-GMPD
Plunger diametre (inches)	2.000	1.625	1.250
Stroke length (inches)		1.30	

Max. working pressure .... 70 (10,000)...............Mpa (psi)

Flow rate Min	25.36(900)	4.5(150)	4.5(150)
m^3/Min (Scf/min)	@ 100 rpm	@ 100 rpm	@ 150 rpm
Max	170(6700)	85(3000)	85 (3000)
	@800 rpm	@ 900 rpm	@ 1200 rpm
RPM maximum	800	900	1200
NPSH required	690(100)	414(60)	345(50)
Kpa (psi)			

2

Water Control

2.1 Introduction to Water Production

In petroleum production, a certain amount of water production is expected and sometimes even necessary in the initial phases of the life of the reservoir or well. A petroleum engineer will have to be able to decide when water control solutions should be applied. If the costs associated with a water production rate still allow for an acceptable operating profit from produced oil or gas, that water production rate is considered acceptable. If the costs associated with a water production rate are too high to allow for an acceptable operating profit margin, the water rate is considered excessive.

Excessive water production can be caused by the natural depletion of a reservoir where an active water drive (either natural or artificial) has simply swept away most of the oil that the reservoir can produce, and there is little left to produce but water. The best completions and production practices can delay, but not stop this water production. Most cases where water-production rates have become a problem could have been avoided or delayed. Understanding reservoir behavior provides a basis for determining whether excessive water production is a concern and to determine if current water production is excessive.

The following issues should be considered when estimating optimum water production rates:

1. Current and projected oil prices
2. A relative cost of high-capacity water handling facilities
3. Cost per volume to dispose of produced water (treatment, transporting, reinjecting, etc.)
4. A relative expense of completing wells to maintain low water production rates
5. Water production needed to produce sufficient oil rates
6. Surface or downhole facilities limited by fluids rate

32 Water Control

7. Water production rate effect on bypassed oil
8. Reservoir maturity
9. Water production rate effect on corrosion rates
10. Water production rate effect on sand production
11. Water production rate effect on scale formation.

2.1.1 Methods to Predict, Prevent, Delay and Reduce Excessive Water Production

1. Oil and Water Production Rates and Ratios
2. Rate-Limited Facilities
3. Water Production Effect on Bypassed Oil
4. Reservoir Maturity
5. Water Production Rate Effect on Corrosion Rates
6. Water Production Rate Effect on Scale Deposition Rates
7. Water Production Rate Effect on Sand Production.

2.1.1.1 Oil and Water production rates and ratios

Operators who own a small percentage of the wells in a field often want to produce the wells as fast as possible, without regard to total reservoir drainage effects. However, fluid production rates need to be controlled because excessive production rates can result in lower ultimate recoveries on a reservoir scale or in a shorter economic lifespan of an individual well.

If a water source exists in the reservoir, oil production rates influence current water production rates and the rate of water production increases. The tools commonly used to predict fluid production ratios and rates for reservoirs and individual wells are:

2.1.1.1.1 Material Mass Balance

Material mass balance calculations will help estimate the total production of the reservoir fluids. These calculations combine the classic concepts of Newton's conservation of mass, the ideal gas law, liquid compression and material solubilities. The derivation and application of this type of technique are described in many reservoir engineering textbooks. The Material Balance Equation can be described as:

> '(Cumulative oil produced and its original dissolved gas + Cumulative free gas produced + Cumulative water produced) − (Cumulative expansion of oil and dissolved gas originally

in reservoir – Cumulative expansion of free gas originally in reservoir) = (Cumulative water entering original oil and water reservoir)'.

The mass balance equation can also be used to predict fluid flow from a well. The cumulative produced fluids would be redefined as well production history and the stock-tank oil initially in place would change to a well-drainage radius.

2.1.1.1.2 Darcy's Law

Darcy's law relates permeability and pressure drop to fluid flow rate. Flow into the wellbore is often considered radial for wells completed in non-fractured zones. Petroleum engineers often perform flooding experiments with reservoir core samples and produced fluids to illustrate the relationship between fluid saturation and the relative permeability of oil and water. This relationship is illustrated in Figure 2.1.

Initial fluid saturations measured after drilling and water-oil rock relative permeability curves (like those in Figure 2.1) can be used to estimate k_{ro} and

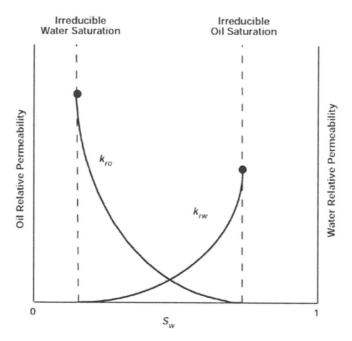

Figure 2.1 Relative Permeability Curves For Water And Oil.

k_{rw}. Then Equations 1 and 2 can be used to predict relative flow rates of oil and water from a zone.

$$q_w = \frac{2\pi k k_{rw} \, h(p_e - p_w)}{\mu_w \, \ln r_e/r_w} \tag{2.1}$$

$$q_o = \frac{2\pi k k_{ro} \, h(p_e - p_w)}{\mu_o \, \ln r_e/r_w} \tag{2.2}$$

In these equations, conditions that have a significant effect on flow rates, well completion type, tubing size and skin damage have been effectively accumulated into the p_w term, rather than considered separately.

The calculations can only provide predictions of relative fluid flow rates and are only useful for the assumed conditions. Saturations, pressures, fluid viscosities, relative permeabilities and skin factors can vary significantly over the lifespan of a well or reservoir; therefore, the relative flow rates calculated with the equations above will also vary over time.

2.1.1.1.3 Productivity index

The productivity index (*PI*) is the ratio of liquid production rate to the pressure drop at the center of the completed interval. *PI* is a measure of a well's potential and can be extrapolated to estimate field potentials. Conditions such as relative permeabilities, skin factors, reservoir pressure and oil viscosity can change throughout the well or reservoir life and can change the *PI*.

2.1.1.1.4 Simulators

Reservoir simulators are tools that use the ability of computers to incorporate concepts such as material mass balance and Darcy's law to predict total reservoir performance. The engineer must be aware of the abilities and limitations of the various types of simulators. Many simulators take into account the varying conditions in a reservoir; however, not all take into account the same variables or assume the same change profiles for these variables. Simulators offer a more convenient way to incorporate reservoir changes when predicting fluid production. Simulator predictions are history-matched with reality to update predictions.

2.1.1.2 Rate-limited facilities

Downhole equipment and surface facilities have a maximum rate at which they can handle fluids. For the downhole equipment, maximum rates generally depend on tubing and orifice sizes, pressure drawdowns (also a function of fluid density) and fluid viscosities.

Factors such as imposed regulations, surface equipment and transport rates can also limit the maximum production rate of a well or group of wells to less than their potential. In the following cases, water production can seriously reduce the oil production rate.

Total water production rates must not exceed the maximum disposal rate. Maximum disposal rates are defined by allowable water discharge volumes, limited separator rates, amount of water that can be transported efficiently from the facility, total water that a pipeline operator allows to flow through an available pipeline and the rate at which water may be reinjected.

An increased water cut can significantly increase the hydrostatic head in the wellbore. This phenomenon will decrease the drawdown pressure at the wellbore and consequently decrease the maximum fluid production rate or stop production entirely.

If the well produces at the maximum rate and water rates increase, oil rates will suffer. In these cases, the oil production decline can be as simple as a one-to-one exchange (one fewer barrel of oil produced for one more barrel of water). However, the change in oil production will be a function of friction pressure and viscosity variations with different water-oil ratios. Bourgoyne (1986) presents a good discussion of these effects.

2.1.1.3 Water production effect on bypassed oil

When water influx is the result of deeper, reservoir-related water production mechanisms, unchecked water production can result in a significant decrease in the total volumes of accessible, mobile oil. Higher water production rates from a zone imply that both the relative permeability to water and the water saturation in that zone is increasing. The higher these parameters are allowed to climb, the more difficult it will be to produce oil from that zone again. For example, where excessive water coning has been allowed to occur, pockets of unswept oil can be left.

2.1.1.4 Reservoir maturity

When an operator initiates production in a field, that reservoir may be a recent find or one that has been produced for several years. Conditions in the reservoir change while it is being produced. These changes include the following:

1. The remaining movable oil in place declines.
2. Principle recovery mechanisms often shift from primary to secondary.
3. Reservoir pressures can drop.

36 Water Control

4. Oil viscosities may increase if pressures decline below the bubble point.
5. Connecting aquifer depths change.

2.1.1.5 Water production rate effect on corrosion rates
Water production rates can significantly affect corrosion rates of downhole and surface equipment. Corrosion rates can be connected to kinetic and erosive/corrosive effects. The rate at which corrosion will occur depends on the concentration of corrosive materials (oxygen, H_2S, CO_2, salts, etc.). The sooner fresh volumes of water containing these corrosive materials come into contact with the metal surfaces of either the downhole or surface equipment, the sooner corrosion can take place (this is not necessarily to imply a linear relationship). Some corrosion products can act as a protective coating against further corrosion (such as low-solubility iron oxides). However, if the flow rate becomes high enough, it can erode the coating from the tubing surface and expose the fresh metal surface to corrosive materials.

2.1.1.6 Water production rate effect on scale deposition rates
Water production affects scale deposition rates in a number of ways. Just as water rates can affect corrosion, if the produced water tends to cause scaling, the faster the water is produced, the faster the scaling deposition. Erosion again affects this process. Extreme friction can help erode scale deposits from the tubing. When a waterflooding program is in place, another consideration is the injection water composition. If scaling is increased when the injection and formation water mix, scaling can be dramatically increased when injection water breaks through.

Several operators inject seawater into their waterflooding programs. The formation water in many of these fields contains barium and strontium, and the seawater contains sulfate. The intermixing of these chemicals will increase the probability of scaling. The water cuts for the fields may be as low as 2%–3%. However, there are cases when additional seawater is produced. Wells that begin to produce the seawater in conjunction with the bottom water are often plugged with scale in a matter of weeks. It is therefore important to know not only how much water can be handled economically, but what the potential water source and water production mechanisms will be.

2.1.1.7 Water production rate effect on sand production
Water can weaken cementitious materials that hold the formation in place, allowing sand production. Zones that produce water may therefore have a lower maximum pressure drop at which sand-free production exists.

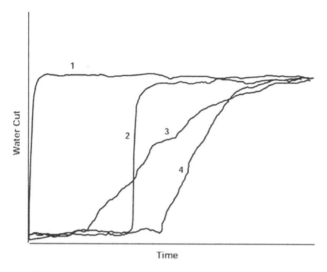

Figure 2.2   Example Water Production History Curves.

2.2 Water Production Mechanisms

Predicting, preparing for and treating for water production involves knowing how that water may be produced. Factors that help determine water production mechanisms include the reservoir drive mechanism, production rates (reservoir and well), connate water and irreducible oil saturations, permeabilities (vertical and horizontal), porosity, permeability anisotropy and heterogeneity, relative permeability/mobility to water and oil, location and continuity of impermeable barriers, reservoir dip, original water-oil contact, a portion of productive interval completed, completion (*e.g.*, perforated, open hole, etc.) location and quality of primary cement job. The commonly observed water production mechanisms and expected water production rates are described below. Figure 2.2 illustrates water-production histories often associated with these mechanisms.

The common mechanisms for water production are:

2.2.1 Completions-Related Mechanisms

2.2.1.1 Casing leaks

Casing leaks can occur in tubing and collars and are often caused by poor completions practices: improper tightening of joints (too loose causing no seal, or too tight causing excess strain), tubing incompatibilities with downhole conditions (temperature, corrosive materials, pressure, etc.). Casing leaks are often observed by a sudden, rapid increase in the water cut (Curve 2 in Figure 2.2).

2.2.1.2 Channel behind casing
Poor cement/casing or cement/formation bonds often lead to channels in the casing- formation annulus. These channels can occur at any time in the life of the well but are usually observed by a rapid increase in water production immediately after a stimulation treatment or unexpectedly high water cut immediately after completion, as in Curves 1 and 2 in Figure 2.2. Channels behind the casing are much more common than casing leaks.

2.2.1.3 Completion into Water
This phenomenon generally occurs when the available data (core data, driller's reports and open hole logs) is either misinterpreted, of poor quality or unavailable. As in some casing leaks, a symptom of completion into water is an unexpectedly high water-cut immediately after production begins, as in Curve 1 in Figure 2.2.

2.2.2 Reservoir-Related Mechanisms
2.2.2.1 Bottomwater
This mechanism is the only commonly occurring water production mechanism that is unavoidable. When a reservoir has an active aquifer driving oil production, as the reservoir depletes water slowly displaces the oil, the wells in the field slowly begin to produce water. The water production history of wells producing water caused by this mechanism even rises if the water table is similar to Curve 3 in Figure 2.2.

2.2.2.2 Barrier breakdown
Natural low-permeability barriers, such as dense shale layers, sometimes separate the oil zone from an aquifer. This barrier can break down for various reasons. If drawdown pressure during production exceeds what the barrier can withstand, it will fail, allowing water to break through and produce. The barrier can also either fracture or dissolve as a result of hydraulic fracturing or matrix acidizing treatments, respectively. A rapid increase in the water production rate can also be an indication of this mechanism. If the barrier is broken during completion (either while drilling or stimulating), a water production history similar to that depicted in Curve 1 of Figure 2.2 would

Figure 2.3 (a) Water Coning (b) Water cresting.

be more representative. If it is caused by pressure depletion or stimulation treatments later in the well life, water production may look more like Curve 2.

2.2.2.3 Coning and cresting

In a water drive reservoir, the drawdown pressure at the wellbore will tend to pull water up into the wellbore. When extreme drawdowns exist in a vertical well, the resulting shape of the near-wellbore water- oil contact is conical; in a horizontal well (Figure 2.3(a)); the shape is more like a crest of a wave (Figure 2.3(b)).

Coning and cresting can be avoided if the well is produced below its critical rate, which is the maximum water-free production rate. Critical rates have been studied extensively. Probably the earliest documentation of critical rate studies was presented in Muskat and Wyckoff (1935). Studies since then have invoked a wide variety of considerations (or conditions) into evaluating critical rates: unsteady states, pseudo-steady states, permeability heterogeneities, horizontal wells and three-phase flow. Most critical rate calculations assume that the rate and cone or crest shape are affected by the ratio of vertical to horizontal permeability, oil zone thickness, the ratio of gravity and viscous forces, well penetration and mobility ratios.

The primary differences in the calculation methods are in the assumptions made to implement simplifications. Muskat and Wyckoff assumed linear flow, whereas Meyer and Garder invoked radial flow. Chierici assumed no influence in a cone shape, whereas Wheatly applied cone-shape calculations. The best method for determining critical rates (or many other values) is the method that most accurately assumes the well conditions at hand. The water production history of a well with a coning problem may look something like Curve 3 in Figure 2.2.

2.2.2.4 Channeling through high permeability

In an ideal, homogeneous, water drive reservoir, the oil is uniformly displaced by the water. However, 'ideal' and 'homogeneous' are rarely applicable to

40   *Water Control*

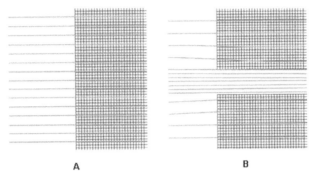

Figure 2.4   FLow pattern of a waterflood through (a) homogeneous permeability and (b) a zone with a high-permeability streak.

reservoirs; there are often layers of varying permeability within a producing interval. As quantified by Darcy's law, the flow rate is faster through higher-permeability layers. The result is a high water production rate through these layers before the water has swept oil from the surrounding layers. In fields where a waterflooding program is in place, this can result in injection and immediate production of injected water with no significant displacement of additional oil. A side view of a 2-D simulation of water flowing from an injector to a producer is shown in Figure 2.4.

In these simulations, 50% of the total pore volume between the injector and the producer was swept with water. Portions (grid blocks) of the rock that were not swept are represented as outlined blocks (•); those that were swept with water have no outline. The bold lines are streamlines representing isoflow.

In the cases where a permeability contrast exists (Figure 2.4b), a significant amount of oil is bypassed by the injection water, compared to the case where no permeability contrast exists (Figure 2.4(a)). A high- permeability layer can result in a rapid rise in the water cut after break-through, as shown by Curve 4 in Figure 2.2.

2.2.2.5 Fracture communication between injector and producer

Natural fractures can provide a direct link between an injector and a producer, allowing the water to flow primarily through these high- permeability channels and bypass oil within the adjacent rock matrix. For highly communicating fracture networks, the water production history may look more like Curve 2 in Figure 2.2, where the abrupt onset of water occurs within a couple of days (or even hours) after injection begins. If the fracture network does not give as direct a path between the injector and producer, the water production history may look more like Curve 4 in Figure 2.2.

2.2.2.6 Stimulation out of zone
In a production well, this occurs when an aquifer is stimulated during a fracturing or matrix acid treatment. In an injection well, this would encompass stimulation treatments that result in decreased sweep efficiency. The effect that this type of problem would have on the water production history is illustrated by Curve 2 in Figure 2.2, where the onset of water would coincide closely with the stimulation treatment.

2.3 Preventing Excessive Water Production

Excessive water production can be treated either in the completion stage or after it becomes a problem. Historically, water production was ignored until it was a real problem—wells often producing in excess of 90% water. Postponing treatments can seriously jeopardize the total well/reservoir productivity, particularly if the water production mechanism is reservoir-related. Medical doctors often promote prevention as the best cure; the same can be said for excessive water production. Simulators can be extremely helpful in deciding which prevention option will be the most successful, both from incremental oil and from an initial investment point of view. Simulators can also help determine if it is worth preventing excessive water production or treating it later. The methods below focus on prevention. Some of the options use chemical treatments.

2.3.1 Preventing Casing Leaks

Casing leaks can be prevented if the tubing is selected that will withstand long-term exposure to the projected pressure, temperature and chemical environments that may exist downhole. Temperature, pressure and chemical resistivities of tubing materials and tubing selection considerations are to be done accordingly.

2.3.2 Preventing Channels Behind Casing

A good primary cement job will usually prevent channels behind the casing. Methods to achieve a good bond between the reservoir and the casing have been established; however, poor primary cement jobs are relatively common.

2.3.3 Preventing Coning and Cresting

Because coning and cresting result from low pressure at the wellbore pulling up the water-oil contact (WOC), techniques to prevent coning and cresting involve ways to minimize the drawdown on the WOC. Holding production rates under the critical rate was the original technique implemented in coning prevention. However, limiting production rates to minimize coning also limits revenue. Other methods to prevent coning involve maximizing the critical rate. Following methods can be used to prevent coning:

2.3.4 Perforating

In the completion stage, the location of perforations has been used to help prevent coning. It is recognized that the farther away the perforations are from the WOC, the lower the tendency for coning will be. The effect of perforation location on the critical rate calculations is illustrated in Equation (2.3).

$$q_{oc} = \frac{4.888 \times 10^{-4} k_H h_o^2 \Delta \rho q_c}{\mu_o B_o} \quad (2.3)$$

Pirson, Meyer and Garder calculated the height of the bottom of the optimum well completion using Equation (2.4). Guo and Lee (1993) presented work that indicates that the completion interval ... should be less than one-third of the total thickness of the oil zone, depending on oil-zone thickness, wellbore radius, and drainage area radius'.

$$h_{cb} = h_o - (h_o - h_p)\frac{\Delta \rho_{og}}{\Delta \rho_{wg}} \quad (2.4)$$

The concept of maximizing the distance between the WOC and the perforations applies to both vertical and horizontal wells. The equations above are designed for application in vertical wells. Calculations for horizontal wells are presented in Chaperone (1986) and Yang and Wattenbarger (1991). This technique is limited to the height of the oil zone. If the zone is thin, the distance between the WOC and the perforations will be small, limiting the completion interval. Another drawback is that the upper interval of an oil zone may have very low permeability.

In completing horizontal wells in the Troll field, Norway shot density was also optimized. The primary objective was to minimize the pressure loss in the wellbore by minimizing the shot densities to reduce the friction factor across the perforated interval.

2.3.5 Fracturing

Fracturing a perforated interval to prevent coning has not been very successful. If designed properly, a hydraulic fracture in a vertical well can help dissipate the wellbore drawdown in a way similar to that of a horizontal well (the water is pulled by a 'line' of pressure, rather than a 'point' of pressure). Generally, vertical permeabilities in hydraulic fractures are very high compared to the matrix horizontal permeability, so extreme care must be taken to ensure that fracture growth does not propagate too close to the WOC.

2.3.6 Artificial Barriers

Placing an artificial impermeable barrier between the WOC and the completed interval will greatly reduce coning tendencies and increase a well's maximum water-free production rate. Artificial gel barriers have often been used to decrease coning, but the vast majority have not been placed until after a water breakthrough has occurred. Problems in waiting come from many sources. In the end, it is very difficult to treat a well for coning and not to plug the entire completed interval. The problem is aggravated by short completion intervals due to thin zones, high K_v/K_h or both.

Success rates have been very high in cases where artificial barriers were placed between the WOC and the completion interval before the wells were placed on production. The general procedure is to (1) drill to the water zone and perforate, (2) place an artificial barrier of a radius sufficient to prevent (or significantly delay) coning and (3) move up the wellbore and perforate into the oil zone.

2.3.7 Dual Completions

Wells can be produced from the water zone and the oil zone to prevent coning. When producing both zones, two production strings are used to keep the oil and water separated— hence 'dual completions'. Dual completions are successful at reducing water coning potentials by decreasing the pressure in the water zone in the wellbore area. Dual completions also reduce the surface handling costs of the produced water by reducing the need for separating the water from the oil.

2.3.8 Horizontal Wells to Prevent Coning

Horizontal wells have been shown to be quite useful in minimizing coning effects. However, cresting still can occur in horizontal wells. Horizontal wells offer the flexibility to disperse the drawdown pressure at the wellbore. In a vertical wall, the water is pulled into the wellbore by pressure concentrated in one spot (the radius of the wellbore), whereas in a horizontal well the pressure is spread out along the length of the wellbore. This is not meant to imply that the pressure is uniformly dispersed down the wellbore; generally, it is found that the drawdown at the heel of a horizontal section is higher than at the toe. In thin oil zones, vertical completion lengths are short, which minimizes the well drainage radius and the amount of oil that can be pulled into that wellbore.

Despite these advantages of horizontal wells, the cost of drilling them is still significantly higher than for a conventional well. (This disparity is declining as the industry improves horizontal well drilling technology). The popularity of horizontal wells has increased over the years because of their increasing capability to prove their worth. Shell has implemented horizontal well completions to reduce coning of both water and gas in the Rabi field, Gabon, etc.

The fundamental concept applied to preventing coning in vertical wells will also apply in preventing cresting in horizontal wells: the drawdown the WOC 'sees' should be minimized. As in a vertical well, water breakthrough times and critical rates are maximized when the lateral portion of the well is placed at the top of the oil zone.

In general, frictional pressure losses cause the pressure drop between the wellbore and the reservoir to be higher at the heel of a lateral than at the toe. This condition is exaggerated in wells that are highly productive, are completed with small tubing or are extended reach and becomes significant if the flowing pressure gradient in the lateral is similar to the producing drawdown pressure. Because of this, water breakthrough from cresting most often occurs at the heel of a lateral. If the drawdown pressure can be better dispersed down the lateral, many cresting problems can be either minimized or delayed. Keeping this in mind, some general statements can be made about completing horizontal wells with significant pressure drop along the lateral: (1) Shot densities should be higher at the toe than at the heel. (2) Stimulation treatments should concentrate more at the toe than on the heel. (3) Impermeable barriers can be placed on the heel of the lateral to decrease cresting tendencies.

2.3.9 Preventing Channeling Through High Permeability

High-permeability streaks can connect a producer either to an underlying aquifer or to an injection well. Minimizing flow through a high-permeability

2.3 Preventing Excessive Water Production

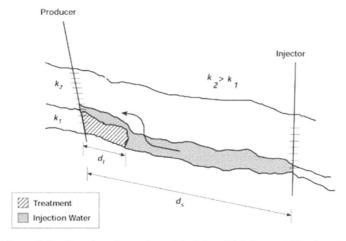

Figure 2.5 Crossflow Bypassing a Block in a High Permeability Streak.

channel is often referred to as profile modification. Preventing flow through high-permeability streaks can involve partial perforating, stimulating or partial blocking. In general, this mechanism can best be prevented if the water flowing through the high- permeability layer is somewhat confined (crossflow minimal) so that water does not flow around blocking treatments or flow into stimulation treatments. Crossflow (illustrated in Figure 2.5) is usually minimized by either a very low-permeability layer adjacent to the high-permeability streak or a low vertical-to-horizontal permeability ratio.

Preventative measures are also successful when the water source and the producer are relatively close: For best results, maximize d_t/d_s (d_t = treatment radius, d_s = distance from well to water source). The most effective way to prevent channeling through a high-permeability streak is to reduce the permeability of the entire streak. However, this usually requires such large treatments that it is economically unjustifiable.

2.3.9.1 Perforating
This option is relatively straightforward. Perforating into the high-permeability streak is minimized (either by density and/or depth) or avoided. This option is particularly successful when crossflow is minimal.

2.3.9.2 Stimulation techniques
Stimulation treatments can help improve the *PI* of a well and disperse the drawdown over the entire perforated interval, thereby decreasing the pressure

46 Water Control

drop across the high- permeability layer. Because the effect of stimulation techniques on preventing channeling through high permeability will be more confined to the near wellbore area, they would be best suited when the water source is relatively close. Possible options include:

1. Stimulating past drilling/completions fluids damage by near wellbore fracturing (High-Permeability Fracturing) or selective HF acidizing. This would be most applicable when the fluids flowing through the high-permeability layer are confined by a low k_v/k_h and when the skin factors in the low-permeability intervals are high enough so that near wellbore stimulation treatments will have a significant effect.
2. Large-scale hydraulic fracturing treatments (Well Stimulation) to increase the effective drainage area of the low-permeability layers: This will be most applicable in cases where the high- permeability layer is sufficiently separated from the interval targeted for fracturing to prevent the fracture growth from entering the high-permeability layer.

2.3.9.3 Permeability reduction

These treatments are injected into the higher-permeability layer(s) to reduce their permeability so that it will be roughly equivalent to the lower-permeability intervals. Treatments of this sort should be only partially, rather than completely, plugging so that the oil in the high-permeability layer is not lost.

If the high-permeability streak can potentially connect an injector to a producer, an engineer must decide which well to treat, if both should be treated or if treatment should be delayed until a breakthrough occurs. If the water flow through the high-permeability layer is confined, options such as a near wellbore stimulation in the producer and avoiding perforating into the high-permeability streak at the injector may be suitable. When there is little chance of preventing the water from finding the high-permeability channel, treating both the injector and the producer may be necessary. The following example illustrates how a simulator might help determine which option would be the most beneficial. In these simulations, 50% of the total pore volume between the injector and the producer was swept with water. Portions (grid blocks) of the rock that were not swept are represented as outlined blocks (•); those that have been swept with water have no outline. The bold lines are streamlines representing isoflow.

This simulation compares the difference in sweep patterns if the injection profile is allowed to go unchecked (Figure 2.6(a)), the injector is treated with a partial blocking treatment an before injection is started (Figure 2.6(b)),

2.3 Preventing Excessive Water Production 47

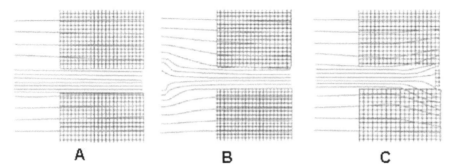

Figure 2.6 (a) Flow Pattern of a Waterflood Through a Zone With a High-Permeability Streak. (b) High-Permeability Streak Treated Before Watering Out: Injector Treated. (c) High-Permeability Streak Treated Before Watering Out: Producer Treated.

or the producer is treated with the same treatment before injection begins (Figure 2.6(c)).

Preventing breakthrough by treating the injector will leave the least amount of unswept oil, according to these calculations.

2.3.9.4 Preventing fracture communication between injector and producer

Again, this water production mechanism is most often treated after water breaks through a fracture network by plugging the fracture at either the injector or the producer with a chemical gallant. This technique can also be used before a breakthrough. Problems with this technique arises when the permeability of the surrounding rock matrix is too low to maintain acceptable production or injection rates after treatment. Other problems associated with this technique have involved poor treatment design. If the agent used to plug the fracture is not strong enough to withstand the drawdown at the wellbore, it may be produced back.

Another technique focuses on well placement. If an operator acquires an older field or a series of wells where it is not economically feasible to drill new wells, this option will be limited. Choosing the location of the injectors relative to the producers is the prime factor, whether in drilling new wells or converting existing wells from producers to injectors. An aerial view of an example well pattern in a reservoir with the indicated fracture orientation is illustrated in Figure 2.7.

In this case, the only well that could be considered as an injector is well A-01. A location of a future injector might be where F-01 is indicated.

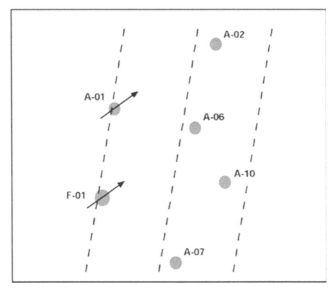

Figure 2.7 Example Well Pattern: Choosing Injectors.

If waterflooding programs already exist, reorienting the direction of the waterflood front is also possible. The McElroy field in the Permian
Basin is an example where this was successful. In this case, the reservoir was naturally and hydraulically fractured. New injection wells were drilled to maximize the injection efficiency, the placements of which were determined by considering the existing wells and the relative fracture orientation.

2.3.9.5 Completing to accommodate future water production rates future zonal isolation

Completing wells with the capabilities to selectively shut off production in a zone (or zones) is a relatively common way to prepare for increased water rates. There are three general ways to achieve this. One method is to build mechanical isolation equipment into the well when it is completed. This procedure is primarily done when excessive water production is expected to enter the wellbore at an interval upstream from the end of the well. Sliding sleeves in combination with packers are commonly used to achieve this effect. As illustrated in Figure 2.8, the sleeves are built into the tubing and packers are placed between the sleeves.

If a higher permeability layer begins to produce water at high rates, the sleeve through which that water is produced can be closed.

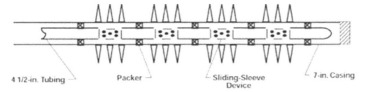

Figure 2.8 Sliding Sleeves Added To Completion To Allow Future Zonal Isolation.

Mechanical zonal isolation is also used to shut off water production from the end of a well. Settable packers such as bridge plugs or slickline settable plugs are tools of choice for this. This procedure is applicable when excessive water rates caused by bottom water are expected. Water would enter the wellbore first at the bottom set of perforations. A slickline settable packer can be placed to shut off the bottom set of perforations and incrementally moved up the hole as the coning progress. Placing plugs such as these is often difficult if the nipple and/or tubing restrictions are small. In the Everest and Lomond fields in the North Sea, a slick mono bore completion was developed that helped maximize these constrictions so that ringless mechanical zonal isolation was achievable.

Planning for selectively injecting chemical plugging agents also involves consideration of the tools required to isolate the target zone (e.g. single packers and straddle packers). Sliding sleeves can also be used when in-depth blocking or wellbore blocking is necessary. Before treatment, all sleeves are closed except that in which treatment will be injected. After the treatment, the closed sleeves are reopened to production and the sleeve that was opened for treatment is closed.

2.4 Creative Water Management

Even when the best engineering techniques for prevention have been implemented, there are cases where excessive water production will be imminent. In other cases, the definition of excessive water production is based on the costs associated with lifting and handling the water. These include costs to replace corroded production strings, mill out upstream carbonate scale (deposition of which is aggravated by release of CO_2), to run an ESP (electrical submersible pump) to offset increased hydrostatic head from an increased height of the water column, to separate oil from water to environmentally safe levels (<40 ppm for the North Sea), and to store and transport produced water.

50 *Water Control*

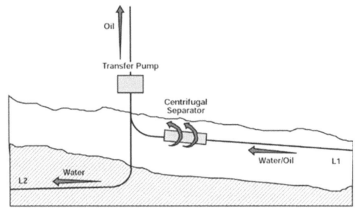

Figure 2.9 Multilateral Well Completed To Accommodate Reinjection Rather Than Lifting of Produced Water.

An option to reduce storage costs, to separate for disposal and to transport produced water is to reinject the water either as a part of a waterflooding program or into a non-communicating reservoir. The cost of drilling enough injection wells and/or converting producers to injectors will at least partially offset the benefits of a reinjection program. Careful planning must be done to accurately anticipate how much water can be reinjected and to avoid typical injection well problems such as scaling and corrosion. Several reinjection programs have been very effective.

Another option that is currently under investigation is to reinject produced water without bringing it to the surface. This option potentially avoids all the costs associated with lifting and handling water. A well completed to handle this might look similar to the one illustrated in Figure 2.9, where multilateral technology is used.

This hypothetical well is located in a reservoir that has a thin oil zone with an active aquifer; therefore, water cresting potentials are high. Lateral L1 is drilled into the pay zone and L2 is drilled into the water zone. Oil and water will be produced from L1 and will be fed into a downhole separator (a device that works like a centrifuge). The oil is then directed to the surface from the separator, and the water is directed down to L2 for subsequent reinjection. The two laterals are directed away from each other to avoid aggravating cresting problems with locally high water pressures caused by injection. Advantages to a design such as this are:

1. Upstream corrosion and scaling are avoided.

2. Oil does not have to be separated as completely from the water as it would if the water were to be disposed of.
3. There will be no pressure reduction due to increased hydrostatic pressure because no water column is formed in the production string.
4. Because the water will be pushed down, rather than up, the strain on the downhole pump will be minimized and its life span lengthened.

2.5 Treatments Used to Reduce Excessive Water Production

Excessive water production is most often treated rather than prevented. The budget available to treat the problem reduces as the water cut increases. This means that little money is available to determine where the water is coming from and why, to pay for the most technically viable solution or to employ proper placement techniques. The keys to the success of shutting off or preventing excessive water are proper problem characterization, appropriate treatment design and effective treatment placement.

2.5.1 Characterizing the Problem

There is no set method for characterizing water-production problems; each case presents a different set of available data for making design decisions. However, the following is a very effective generic procedure.

1. Using all available data on a candidate, assign/calculate values to as many of the factors that help determine water- production mechanisms as possible.
2. List the unknowns and how they can be determined. For example, PLT or downhole video can help pinpoint where water is entering the wellbore
3. Compare the risk associated with designing a treatment with the given knowns *vs.* the cost of collecting more data. For instance, if an engineer cannot distinguish between a channel behind the casing or a high-permeability streak without a cement bond log, the engineer must determine if it is worth the risk of treating for the wrong mechanism, rather than spending the money to run the log. Often treatment volumes for high-permeability streaks are higher (and therefore more expensive) than for channels behind the casing. If the engineer chooses to squeeze cement to fill a channel behind the casing when the problem was really a high-permeability streak, he or she may stop water production for a short time, but additional treatment may be necessary shortly after the

squeeze treatment. On the other hand. The cost to run the log must also be considered. If the candidate well is an off-shore sub-sea completion well, the cost to run the log may be several times more than the cost of either treatment.
4. Estimate post-treatment production potential.

2.5.2 Treatment Design

Whether preventing or shutting off excessive water, there are two primary questions that must be answered for proper solution designs: (1) What is the treatment expected to do? (2) What conditions must the treatment withstand? This section will focus on shutting off excessive water; however, many of the concepts are transferable to preventing excessive water production.

2.5.3 Expected Treatment Effect on Water Production

Success must be realistically defined for any operation to be successful. This sounds simplistic; however, there are many cases where an operator does not fully communicate what results are expected of a treatment or recognize expectations that are unrealistic. It may not be possible to decrease the water cut to 0% and to increase oil production by a factor of 10. The same information used for identifying the water production mechanisms and to determine the maximum amount of water is also used to realistically define what treatment is expected to do.

As an example, enough information was available to determine that Well A-01 is producing excessive water through the high-permeability interval, Z1, from an underlying aquifer and no water production from zones Z2 or Z3 (Figure 2.10).

The potential for vertical communication between Z1 and the adjacent zones is known to be minimal, but the water saturation of Z3 was 35% when the well was new (and at least that currently). If production from Z1 is completely shut off, water will likely begin to produce from Z3. Although this zone was not contributing to the water production prior to treatment, the producing pressure that Z3 sees would increase due to fewer perforations open and possibly a shorter water column. Assuming Z1 is shut off, an operator can predict what the new production index and water rate will be. The operator can also calculate the new critical rate for the well to determine if it is worth keeping the production rate low enough to prevent additional water production due to coning.

2.5 Treatments Used to Reduce Excessive Water Production

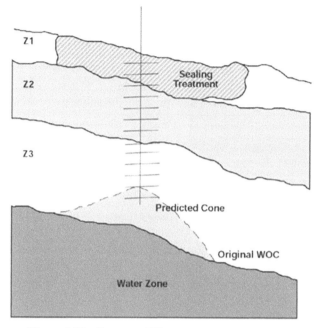

Figure 2.10 Treatment Effect on Example Well A-01.

In this example, the treatment was expected to completely shut off production in the target interval; however, complete seals are not always advantageous. Whenever water permeability is decreased, the water drive mechanism displacing oil is decreased. Generally, treatments can only be effectively placed a short distance from the wellbore—rarely more than 50 ft, a very small area on a reservoir scale. Although water shut-off treatments may not significantly affect the water drive mechanism on a reservoir scale, cutting off a well from the water source completely can have serious detrimental effects on the well's productivity. Also, the target interval may still contain a significant volume of mobile oil.

2.5.4 Treatment Types

2.5.4.1 Zone sealants

Definition: When a zone sealant is properly applied, all production of the treated zone is shut off. There are two classes of zone sealants: wellbore and matrix sealants. Wellbore sealants can be either chemical or mechanical. The chemical sealants can either be strong gallants or cement and are designed not to appreciably penetrate the rock matrix. Mechanical sealants are generally packers or sliding sleeves that are either built into the completion or run into the well after excessive water production occurs. Matrix sealants are chemical treatments that are injected into the rock matrix of the target zone and subsequently reduce the absolute permeability of the treated rock to zero. The size of any wellbore sealant must be sufficient to cover the entire interval.

Mechanism: A zone sealant must be capable of completely plugging all flow channels that connect the reservoir to the wellbore. Mechanical and chemical wellbore sealants rely on the strength of the bulk material and the bond between the material and the tubing; weakening of either can cause the seal to fail. Matrix sealants must also be resistant to downhole environments. Because the materials are injected farther into the rock matrix, if some degradation or syneresis of the treatments occurs, the seal can still remain intact (Bryant, 1996). The paths through which fluids flow through matrices comprise pore spaces connected by pore throats, rather than bundled capillary paths. Because of this, matrix zone sealants do not need to plug every pore space and throat, just enough of them so that there are no open flow channels connecting the reservoir to the wellbore.

Lifetime: Generally, sealants are expected to last indefinitely, or at least until the economic benefit or the treatment sufficiently offsets its cost.

2.5.4.2 Permeability-Reducing Agents (PRA)

Definition: A PRA must be able to reduce the water production from the target interval. Most PRAs are matrix treatments, but sand plugs placed in the wellbore can also be designed to reduce, rather than the plug, permeability. A PRA can also reduce the oil permeability of the target zone.

Mechanism: A PRA that decreases the permeability by 50% can work either by completely plugging half of the flow channels in a rock matrix, by partially blocking all the flow channels or by completely plugging less than half of the flow channels and partially blocking others. Materials that are used for zone sealants may also be designed to partially reduce permeability by adjusting formulation, placement, volume or a combination thereof.

Lifetime: Generally, sealants are expected to last indefinitely, but the ability of a matrix sealant to maintain its ability to reduce water rates can be a function of the zone water saturation.

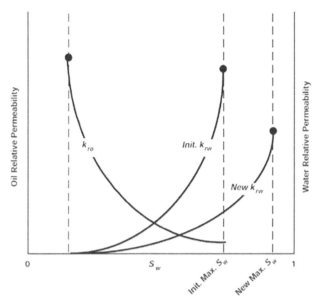

Figure 2.11 How a Relative Permeability Modifier Might Adjust Relative Permeability Curves.

2.5.4.3 Relative Permeability Modifiers (RPM)

Definition: The true definition of an RPM has been the subject of a great deal of debate. In the purest sense, a material that reduces the relative permeability to water more than to oil is considered to be an RPM. Some argue that if a treatment will decrease the water-oil ratio (WOR) that is produced from the target zone, it should also be considered an RPM. Both definitions imply that when an RPM is applied, the relative water-oil producing rate will decrease. The first definition also requires an RPM to shift the relative permeability vs. saturation curves (Figure 2.11) so that when residual oil saturation is reached in the rock matrix (at maximum water saturation), the k_w/k_o is lower.

Lifetime: The issue of RPM lifetime has also been debated. Chemically, RPM lifetimes are defined by how long it takes for the treatment to produce back. Engineering studies indicate that when a material reduces the WOR of an interval, eventually the water driving production will build upon the other side of the treated area. When the water saturation of the outside area rises to match the original matrix maximum water saturation, the oil permeability will go to zero: no more oil will be produced from that interval. Simulations can predict how long a scenario of this type will take. When an RPM loses effectiveness in field applications, a retreatment can help determine if the original treatment failed because it was producing back slowly throughout its lifetime, if it was allowing water saturation buildup outside of the treatment area or if another reason caused it to fail (e.g., water breakthrough from a new water-producing mechanism). If the second treatment is successful, it is likely that the first treatment failed to remain in the rock matrix.

Mechanism: A third issue that still has to be resolved is the mechanism or mechanisms by which RPMs reduce water permeability more than oil permeability. Selective pore throat plugging and selective rock surface effects are among the most postulated mechanisms.

Selective pore-throat plugging implies that the disproportionate permeability effects rely on the ability of a hydrophilic RPM to selectively invade and subsequently plug more water-saturated pore spaces. Laboratory data indicate that some RPMs do, at least partially, rely on this mechanism. RPM formulations that are designed to form a weak gelatinous material after they are placed in the rock matrix will likely rely the most on this mechanism.

It is also postulated that some RPMs adsorb to the surface of the rock and selectively interact with the water. This selective interaction may include lubrication or hydrophilic film effect, where the RPM increases the relative oil mobility or a gel shrinking and swelling effect where the polymers tend to shrink more in oil than in water, causing the pore sizes to be smaller when more water is present. An RPM system that is designed to be adsorbing, but will not necessarily thicken in the rock matrix, will probably rely on this mechanism.

2.5.5 Description of Previously Applied Treatments

The applications of currently available systems to either promote an effect or to treat a specific water production mechanism are listed in Table 2.1. Brief descriptions of these systems are provided below.

2.5.5.1 Mechanical plugs

These include mechanical tubing packers and sliding sleeve devices. These can be built into the tubing and inflated or closed by wireline or slickline devices. Mechanical packers can also be run into the well and set after the well is completed. Mechanical water shut-off devices have been used successfully throughout the world, and are most useful when there is little potential for the water to flow to another open section of the well. Mechanical plugs can be used to seal any section of the completed interval.

2.5 Treatments Used to Reduce Excessive Water Production   57

Table 2.1 Systems Used to Affect Permeability and to Treat Specific Water-Production Problems

Desired Treatment Effect	Mechanical Plugs	Sand Plugs	Water-based Cements	Hydrocarbon-based Cements	Internally Actiuatet Silicates	Externally Activated Silicates	Monomer Systems	Cross linked Polymer Systems	Wettability Modifiers	Microbial Systems	Foams
Zone Isolation	X		X	X	X	X	X	X			
Permeability Reduction					X		X	X		X	X
Relative Permeability Modification								X	X		
Problem											
Acidized into water	X			X	X		X	X			
Barrier break down			X	X	X	X	X	X			
Bottom water	X	X	X		X	X	X	X			
Casing leaks			X	X	X	X	X				
Channel behind casing			X	X		X					
Coning or cresting					X		X	X	X	X	
Fractured into water				X	X	X	X	X			
channel from injector			X	X	X		X	X	X		
channel to producer					X		X	X	X		
into aquifer	X		X	X	X		X	X	X		

2.5.5.2 Sand plugs

Sand can be placed in the wellbore to reduce or shut off production of the lower interval of a well. Sand sieve size and plug size can be adjusted to cause either a total seal or a partial plug. Often when a total seal is required, the last portion of the plug can be mixed with cement (or another chemical sealant) to reinforce the seal.

2.5.5.3 Water-based cement

These include either standard or ultrafine cement slurried in water. The small particle size of the ultrafine cement can allow more complete penetration into micro-channels.

2.5.5.4 Hydrocarbon-based cements

These also include both standard and ultrafine cement, but are slurried in oils (usually diesel). The slurries contain surface-active materials that allow them to absorb water from an external source. Oil-based cement are designed to be placed anywhere in a well, but only set if they come into contact with water, allowing a certain amount of selectivity.

2.5.5.5 Externally activated silicates

These systems generally comprise two stages. The first stage contains a material that will gel a silicate instantaneously upon contact (usually a water-thin $CaCl_2$ brine). The second stage contains the silicate source (viscosity varies from one similar to honey to that of water, depending on the concentration of silicate). These silicates are pumped with inert spacers between the stages to keep them separate until they reach the target area. The resulting gels are stiff, brittle solids.

2.5.5.6 Internally Activated Silicates (IAS)

These systems are generally placed as water-thin freshwater based solutions: a silicate source and an activator designed to trigger gelation of the silicate at a predesignated time. The gel times of silicates depend on the system pH and temperature. Gel times of most currently applied IAS systems are controlled by pH, taking the downhole temperatures into account. The target pH is either achieved on the surface by strong or weak acids or in situ by materials that slowly degrade (either thermally or with time) to form acids. The resulting gels are stiff, brittle solids. Although their effective permeability reduction in a rock matrix can decrease with silicate concentration, they tend to reduce oil and water permeability equivalently. IAS systems have been very effective in field applications.

2.5.5.7 Monomer systems

These systems are placed as water-thin solutions containing a low molecular weight material (monomers or oligomers) and an activator. After placement, the activator initiates the polymerization of the monomeric or oligomeric material and results in a solution with a much higher viscosity. Polymerizations are usually activated by adjusting the system to a pH that will allow polymerization at the required time at downhole temperatures (similar to the IAS systems), or by the slow decomposition (either thermal or with time) of the activator to form free radicals capable of initiating polymerization. Monomer systems that have been used commercially include (1) phenol and formaldehyde controlled with pH, solid gels; (2) resorcinol and formaldehyde controlled by pH, stiff fragile gels; (3) acrylamide and an optional bisacrylamide-crosslinker activated by the decomposition of a time-delayed oxidizer, molasses-like to rigid ringing gels and (4) bifunctional amino acrylate initiated by thermal decomposition of an oxidizer, lipping to rigid ringing gels. The ability of these systems to reduce permeability is relative to monomer concentration. They also seem to have some relative-permeability effects at low concentrations.

2.5.5.8 Crosslinked polymer systems

These are polymer and crosslinker-containing systems that are placed in the rock matrix at viscosities low enough to allow injectability (generally, 10– 200 cp: injectability will depend upon formation permeability). After placement, the systems crosslink to form thick viscoelastic gels (from lipping gels to rigid gels, while the concentration and crosslinking are increased). At high concentrations, these systems are used for zone isolation. At lower concentrations, they have been used for either permeability reduction or relative permeability modification.

The polymers used are normally water-soluble: partially hydrolyzed polyacrylamides (PHPA), thermally stabilized copolymers of PHPA, non-hydrolyzed polyacrylamide (NHPA), cationic polyacrylamide, polyvinyl alcohol, guar, guar derivatives, xanthan and scleroglucan. PHPA and its copolymers are most commonly used; biopolymers have rarely positive results. Most of the polymers start out cross-linkable, so they need to rely on the crosslinker chemistry for the delay. NHPA has no place for the crosslinkers to attach, so its crosslinking is delayed by the slow hydrolysis of the polymer to form crosslink sites.

Metallic and organic crosslinkers have been used, both of which are generally pumped as 'masked' materials that are unable to interact with the polymer until their masks are removed. Metallic 'masks' are called ligands which are strongly attracted to the metal ion by ionic forces. The stronger this attraction and larger the ligand, the longer it takes for the metal to release to crosslink the polymer. The rate at which the metal is released can be controlled by pH or by the ligand concentration in the system. The Excess ligand can be added to the crosslinker or the polymer/crosslinker solution to delay the metal release, but this often results in weaker crosslinking interactions of the metal with the polymer. Metallic crosslinkers used commercially include chromium acetate, chromium propionate, zirconium lactate and aluminum citrate.

Organic crosslinkers work in one of two ways: (1) a weakly-attached organic group that is connected to the part of the crosslinker molecule that would crosslink slowly hydrolyzes off, leaving the crosslinker molecule free to react with the polymer. (2) components that can slowly form the crosslinker are added to the polymer solution, rather than a crosslinker.

2.5.5.9 Surface-active RPMs

These systems are generally pumped as low-viscosity solutions containing materials (generally polymers) that adsorb to the surface of the rock matrix. When in place, these materials primarily interact with the rock surface; no significant crosslinking or polymerization is expected (further explanation is given above in the RPM section). Materials that have been used for this include PHPA, amino acrylate copolymers and surfactant-alcohol blends.

2.5.5.10 Foams

Foams have been used to reduce water production, but not often. Generally, these materials are placed as solutions with either dissolved gas that expands after placement or with no gas so that it subsequently foams with contact with the gas downhole. Foams are more commonly used for more in-depth mobility control but can be used as blocking agents as well.

2.5.6 Treatment Lifetime

In general, treatments are expected to remain effective at reducing or shutting off water production under downhole conditions indefinitely, depending on the stability of the treatment and the predicted water-production rates in the months and years following the treatment. For treatments that completely shut off production of the target interval, the treatment should prevent further production from that interval indefinitely. For treatments that are designed only to reduce the permeability of a zone (reducing either the absolute permeability, the ratio of oil and water permeabilities or both), the ability of the treatment to maintain low water rates after placement will change as the relative saturation of water changes. In some cases, the relative water saturation will remain constant, but more often, water saturations of a given producing interval increase with the age of that interval. As the relative permeability curves in Figure 2.1 indicate, the relative water permeability of that interval will also rise if the treatment will allow it. Studies have predicted that if treatments force the relative water permeability of an interval to remain constant even as water saturations rise, a water block may result in that shuts off all production from that interval.

For many treatments that are to be injected into the rock matrix, the ability of the treatment to reduce permeability relies on the material's resistance to degradation at downhole conditions. This is often tested in the laboratory by exposing bulk treatment samples to reservoir fluids and temperatures. Stability is then judged by a treatment's ability to maintain its initial bulk properties. Most chemical gallants are often only considered stable if the gel viscosity remains constant and if it does not synerese (shrink; leaving partially dehydrated gel balls in water that was originally a part of the bulk gel).

If a gel is exposed to reservoir fluids and temperatures and maintains its bulk viscosity with no apparent syneresis, it will probably survive in the rock matrix. The opposite cannot be assumed: if a bulk gel sample degrades or synereses when exposed to reservoir fluids or temperatures, it may still maintain all of its ability to reduce/stop water flow in the rock matrix (Bryant et al., 1996). The reason for this behavior has not yet been defined. Although bulk testing is relatively convenient, more accurate stability determinations are performed by flooding core material with candidate treatments and monitoring the treatments' long-term effectiveness. It is not completely understood why there is this disparity. However, if the bulk treatment degrades but maintains effectiveness in a core, it is possible that gel in the rock matrix did not degrade because it was exposed to less oxygen (oxygen can be very effective at enhancing polymer degradation). If the bulk gel synereses when exposed to reservoir fluids and temperatures but maintains effectiveness in the rock matrix, it is possible that the gel does not syneresis in the rock matrix or that after syneresis, the remaining concentrated gel particles block pore throats in such a way that no open channels through which fluids can flow exist.

2.6 Selecting Treatment Composition and Volume

Choosing which treatment to use to either reduce or prevent excessive water production is an exercise in balancing technical aspects (strength, depth and stability requirements) with economic aspects (volumes, concentrations and treatment types). The best technical solution may also be twice the cost of the next best option, but may potentially result in three times the economic benefits. On the other hand, an inexpensive, technically inferior treatment may not reduce water production enough to offset even its minimal cost. Some general relationships can be drawn: strength is proportional to concentration and depth and volume requirements decrease with increasing gel strength. Another important factor is environmental regulations. When a system is chosen, it must also meet the environmental and toxicity requirements of the area of application.

When the expectations of treatment are defined and the placement and long-term downhole conditions are determined, available treatments are screened for their capability to fulfill these needs. This screening, along with matching environmental legislation, may result in one choice or several. It may be determined that there are a few candidate systems either based on different chemicals or based on the same chemicals but varying in concentration or penetration requirements. For instance, a crosslinked polymer treatment made of the same chemical components will seal the target zone if Formulation A is used, and will only reduce the target zone's permeability if Formulations B or C are used (Table 2.2).

Formulation B is an example of a lower concentration system that will produce a gel that causes less damage to a given volume of rock than Formulation A; Formulation B will result in a weaker gel. Formulation C can be just as damaging to the rock in which it is injected, but there is less of it to create the effect. This example can only apply if Formulations A and C are of low enough concentration that their ability to create damage is relative to their penetration depth (some systems are so concentrated that they need no penetration to form a seal).

Table 2.2 Comparison of Treatments

Formulation A: Zone Sealant	Formulation B: PRA	Formulation C: PRA
1.5% Polymer A	0.7% Polymer A	1.5% Polymer A
1000 ppm Crosslinker X	500 ppm Crosslinker X	1000 ppm Crosslinker X
Mixed in 2% KCl brine	Mixed in 2% KCl brine	Mixed in 2% KCl brine
Treatment Volume: 100 m^3	Treatment Volume: 100 m^3	Treatment Volume: 40 m^3

2.6.1 Placement Techniques

2.6.1.1 Bullheading

This is a single injection method (one fluid pumped at a time) where the treatment is pumped down the existing tubing with no mechanical zone isolation to divert it into the target zone. This can be an effective method if the only open zone is the target zone, or if the treatment will not significantly damage zones adjacent to the target zone. Bullheading is often the least expensive and most simple placement technique.

2.6.1.2 Mechanical packer placement

This is also a single injection technique that uses a packer to isolate the target zone. This includes isolating with straddle packers, single packers, sand plugs or a combination.

2.6.1.3 Dual injection

This technique is a hybrid of the mechanical packer technique. Mechanical isolation of the target zone is first achieved. To prevent invasion of the treatment into an adjacent zone, a second non- damaging fluid is pumped simultaneously. If the target zone is above the zone that needs protection, the treatment is placed down the tubing annulus and the second fluid down the tubing; if the target zone is below the zone to be protected, the treatment is placed down the tubing and the second fluid is placed down the annulus. The injection pressure of both fluids is kept equivalent.

2.6.1.4 Isoflow

This placement technique provides a method of directing the treatment to the target zone without using mechanical zone isolation. The end of a tubing string is placed at the top of the target interval. Two fluids are pumped simultaneously, one down the tubing and the other down the annulus, depending on the relative location of the target zone. A logging tool is hung at the end of the tubing. The fluid that is pumped down the annulus is doped with a material that can be detected by the logging tool. During placement, the location of the interface between the two fluids is monitored and the pump rates are adjusted throughout the job so that the level of this interface remains at the end of the tubing. When the logging tool detects the material going down the annulus, the relative pump rate of the fluid going down the tubing is increased. The following two considerations have to be looked into:

2.6.2 Viscosity Considerations

When zone sealing or absolute permeability reduction fluids are needed, zonal isolation to prevent damage of non-target intervals (or to prevent unnecessary waste of treatment) is recommended. In some cases, however, the target zone cannot be completely isolated or is internally laminated with varying permeability lenses so that it is not easy (or possible) to isolate the fraction of water producing perforations.

Systems that do not completely seal the zone are often chosen for this type of application. In this case, it might be most effective to treat with a sealing formulation that will preferentially penetrate in the higher-permeability, watered-out lens. Studies have indicated that materials that are placed as water-thin (or very low viscosity) tend to preferentially inject into the higher permeability layers; higher viscosity fluids tend to self-divert. The risk of damaging adjacent lower- permeability lenses can also be reduced by careful job monitoring and/or variations in treatment formulation throughout the job.

2.6.3 Temperature Considerations

For systems that react after placement, treatment compositions also often determine set times. When treating with systems that gel in the rock matrix, wells are shut in after placement until the treatment thickens to prevent placed treatments from either being produced back, flowing into a non-target zone by crossflow or being pushed away from the wellbore by injection water. Historically, treatment gel times were designed based on the placement time required (pump- rate dependent) and the bottom hole static temperature (BHST). This assumes the treatment will be at BHST while it is set, an appropriate assumption for low-volume treatments. However, when fluids are injected into a zone, the temperature of that zone cools down significantly: Cooldowns are more dramatic at higher injection rates and volumes.

Simulations indicate that when wells are shut in after treatment, the time required for a zone to reheat to reservoir temperature can be days or even weeks (Figure 2.12).

If a large treatment is required, cooldown simulations should be performed and treatments should be designed to react in a reasonable time frame. This can mean either adjusting the pumping rate, the treatment formulation or both. The decision can also depend on the system chosen; there may be temperature/time limitations or time/strength dependency. For instance, if an IAS system is to be placed, it might be better to inject the treatment at as high a rate as possible and adjust the activator concentration or composition so that

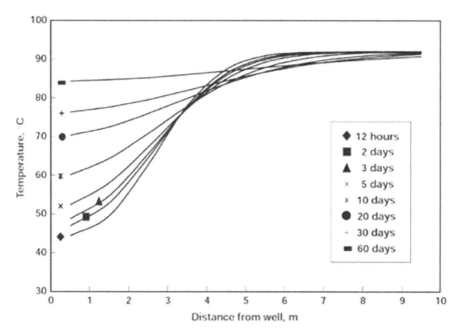

Figure 2.12 Predicted Heat-Up Profile of a Treated Well: Treatment Injection Time = 12 Hr, BHT = 91°C, Treatment Radius = 7 M.

the material placed last will be set within a couple of hours at the predicted low temperature, rather than to place a slow-setting IAS formulation slowly to avoid cooldown. The former will result in a stronger gel at the wellbore, where needed, and a shorter placement time (shorter down-time of the well, less loss of revenue and faster treatment pay out).

References

Ahmed, U. (1991). "Horizontal Well Completion Recommendations Through Optimized Formation Evaluation," paper SPE 22992.

Bale, A., Smith, M.B., and Settari, A. (1994). "Post-Frac Productivity Calculation for Complex Reservoir/Fracture Geometry," paper SPE 28919..

Beggs, H.D. (1991). "Production Optimizing Using Nodal Analysis," Oil and Gas Consultants Int. Inc.

Bryant, S.L., Rabaioli, M.R., and Lockhart, T.P. (1996). "Influence of Syneresis on Permeability Reduction by Polymer Gels," paper SPE 35446.

Chaperone, I. (1986). "Theoretical Study of Coning Toward Horizontal and Vertical Wells in Anisotropic Formations: Subcritical and Critical Rates," paper SPE 15377, 1986.

Dake, L.P. (1994). *The Practice of Reservoir Engineering*. The Netherlands: Elsevier.

Dikken, B.J. (1990). "Pressure Drop in Horizontal Wells and Its Effect on Production Performance," paper SPE 19824.

Gilman, J.R., Bowzer, J.L., and Rothkopf, B.W. (1994). "Application of Short-Radius Horizontal Boreholes in the Naturally Fractured Yates Field," paper SPE 28568.

Harrison, R.D., Restarick, H., and Grigsby, T.F. (1994). "Case Histories: New Horizontal Completion Designs Facilitate Development and Increase Production Capabilities in Sandstone Reservoirs," paper SPE 27890.

Kortekaas, T.F.M. (1985). Water/oil displacement characteristics in crossbedded reservoir zones. *SPEJ* 1985, 917–926.

Lake, L.W. (1989). *Enhanced Oil Recovery*. Hoboken, NJ: Prentice Hall.

Lockhart, T.P. and Albonico P. (1992). "A New Gelation Technology for In-Depth Placement of Cr^{3+}/Polymer Gels in High-Temperature Reservoirs," paper SPE 24194.

Maddox, S., Gibling, G.R., and Dahl, D. (1995). "Downhole Video Services Enhance Conformance Technology," paper SPE 30134.

Mumallah, N.A. (1988). Chromium (III) propionate: A crosslinking agent for water-soluble polymers in hard oilfield brines. *SPERE* 1988, 243–250.

Nolan-Hoeksema, R.C., Avasthi, J.M., Pape, W.C., and El Raba, A.W.M. (1994). "Waterflood Improvement in the Permian Basin: Impact of In-Situ-Stress Evaluations," paper SPE 24873.

Richardson, J.G., Sangree, J.B., and Sneider, R.M. (1987). "Coning," paper SPE 15787.

Schilthuis R.J. (1936). Active oil and reservoir energy. *Trans.AIME*.

Stavland, A. and Jonsbraten, H.C. (1996). "New Insight Into Aluminum Citrate/Polyacrylamide Gel for Fluid Control," paper SPE 35381.

Sydansk, R.D. (1993). "Acrylamide-Polymer/Chromium (III) – Carboxylate Gels for Near Wellbore Matrix Treatments," paper SPE 20214.

Turki, W.H. (1985). "Drilling and Completion of Khuff Gas Wells, Saudi Arabia," paper SPE 13680.

van Batenburg, D.W. (1991). "The Effect of Capillary Forces in Heterogeneous 'Flow-Units': A Streamline Approach," paper SPE 22588.

Vidick, B., Yearwood, J.A., and Perthuis, H. (1988). "How to Solve Lost Circulation Problems," paper SPE 17811.

Vinot, B., Schechter, R.S., and Lake, L.W. (1989). "Formation of Water-Soluble Silicate Gels by the Hydrolysis of a Diester of Dicarboxylic Acid Solubilized as Microemulsions," paper SPE 14236.

Wilkie, D.I., Kennedy, W.L., Tracy, K.F. (1996). "Produced Water Disposal—A Learning Curve in Yemen," paper SPE 35030.

Yang, W. and Wattenbarger, R.A. (1991). "Water Coning Calculations for Vertical and Horizontal Wells," paper SPE 22931.

Zaitoun, A. and Kohler, N. (1988). "Two-Phase Flow Through Porous Media: Effect of an Adsorbed Polymer Layer," paper SPE 18085.

3

Sand Control

3.1 Sand Control Introduction

The production of formation sand with oil and/or gas from sandstone formations creates a number of potentially dangerous and costly problems. Loses in production can occur as the result of sand partially filling up inside the wellbore. If the flow velocities of the well cannot transport the produced sand to the surface, this accumulation of sand may shut off production entirely. If shutoff occurs, the well must be circulated or the sand in the casing must be bailed out before production can resume.

Once produced sand is at the surface and no longer threatens to erode pipe or reduce productivity, the problem of disposal remains. Sand disposal can be extremely costly, particularly in offshore locations where environmental regulations require that the produced sand must be free of oil contaminants before disposal.

Subsurface safety valves can become inoperable, leading to large economic losses and personal hazards, particularly at offshore and remote locations. Erosion-damaged surface and subsurface equipment (Figure 3.1) is expensive to replace and valuable time is lost during replacement and repair.

3.1.1 Formation Damage

This is another problem associated with wells that produce sand unchecked. The possible creation of void spaces behind the casing can leave the casing and any shaly streaks in the reservoir unsupported. Specifically, the casing can be subjected to excessive compressive loading, causing collapse or buckling. The much less permeable shaly streaks that remain can collapse around the perforated casing, causing severe and irreparable restrictions to production. Failure to prevent formation sand production in its early stages can therefore be very expensive in terms of eventually lost revenue and

70 Sand Control

Figure 3.1 Surface Valve, Eroded by Sand Production.

additional operating costs. In addition, it can create potentially hazardous conditions at the well site.

3.1.2 Fines Migration

Tiny solid particles occur in the pore spaces of all sandstone formations. Fines are usually identified as any solids that pass through a 400 mesh (37 microns) screen, which is the smallest screen size available. Electron micrographs have revealed the nature of typical fines – clays, quartz, minerals (such as feldspars, muscovite, calcite, barite) and amorphous material. Clays are only a small constituent. Quartz and amorphous material make up most of the fines.

Fines are loose, not attached to the sandstone grains and can move freely within the porous matrix. Experiments have shown that fines movement is strongly affected by liquid phases and boundaries. Fines are only thought to move readily when the phase that wets them is flowing. Mixed wettability fines are confined to the oil/water interface.

It has been suggested that imposing a high initial drawdown (opening the chokes rapidly) causes formation damage due to fines migration. Wells which are completed and brought online with minimal pressure surges to the formation should have lower skins. In some cases, large under-balanced perforating may lead to reduced near-wellbore permeability due to fines migration caused by the under balance surge. Furthermore, in some fields rapid productivity declines are observed over periods of a few months. Some of this is attributed to the movement of formation fines during normal production from the well.

3.1 Sand Control Introduction 71

Fines cause productivity impairment when they bridge across pore throats. Bridging is a result of two things:

1. Fines move only when the fluid velocity is sufficiently high to entrain the particles. Thus fines migration only occurs in a cylindrical region of a few feet around the wellbore.
2. Mobile fines accumulate, bridging pore throats and impairing productivity, when they are of suitable size.

FINES MIGRATION

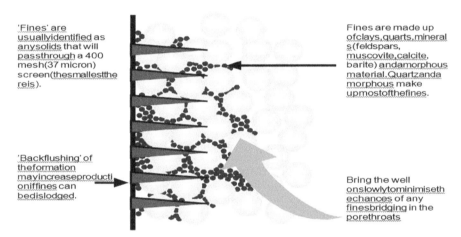

Laboratory tests indicate that changing the flow direction can increase permeability, at least temporarily. This is as a result of fines bridges across pore throats breaking up. Such observations have to lead to the technique of backflushing a well, a useful technique for the identification of fines migration. Some wells respond simply to injecting a filtered fluid into the formation for a few hours (below fracture pressure), which breaks up many of the fines bridges. This results in higher flow rates after the treatment but does not prevent the onset of a further decline as the fines continue to migrate.

3.1.3 Sand Production Mechanisms

Before sand exclusion measures can be optimized, the following sand production mechanisms must be understood:

1. **Grain-by-Grain Movement.** Sand movement away from the formation face probably accounts for most formation failures. A well with this type

of formation rock may eventually require some type of sand control. If the well is allowed to produce sand for an extended period, the choice of sand control method may become more limited.
2. **Movement of Small Masses.** Under some circumstances, formation rock can break away, resulting in rapid failure. Generally, this type of failure will result in a sanded wellbore that will not produce after sand covers the perforations.
3. **Massive Fluidization.** The massive amount of sand produced either prevent production or causes erosion; sand disposal problems become too great for production to continue.

Most recommend that sand control techniques be applied immediately upon indication that a formation will produce sand. This practice will allow the highest success rate and the lowest production loss possible after sand control is applied.

Laboratory studies have shown that once unconsolidated sand is disturbed, the sand cannot be packed back to its original permeability. Therefore, sand control should be applied before the reservoir rock is seriously disturbed by sand production. Regaining the original density and porosity is not difficult, but the permeability will always be much less than the original permeability. Therefore, if a well is allowed to produce enough sand to seriously disturb the reservoir rock around the wellbore, restressing the reservoir during sand control application can result in a lowered permeability zone around the wellbore.

Wells that have no sanding tendencies can be completed in a number of conventional ways. If no initial indication of sand production exists, but it could become a problem in the future, retrievable packer equipment should be used during the completion, so that work overs can be performed easily if sand control treatments become necessary.

3.2 Formation Sand

Sand can be defined as a loose, unconsolidated, granular material. Sand is of the size range 0.06325–2mm. the chemical composition of sand is as follows:

1. Quartz: 78.66%
2. Al_2O_3
3. CaO
4. Fe_2O_3

5. MgO

Among these quartz is most stable. Sand is mildly hydrophilic in nature.

3.2.1 Petro Physical Properties

1. Porosity:
    1.1 Decreases with poor sorting, tighter packing and more cementing.
    1.2 Unchanged by differing grain size.
2. Permeability:
    2.1 Decreases with poor sorting, tighter packing and smaller sand grains.
    2.2 Often directional in sandstones and highest along a bedding plane and direction of the depositional flow.
    2.3 Horizontal permeability is greater than vertical permeability.
    2.4 Thicker beds have higher permeability.
3. Oil – Water Wet
    3.1 Sands mostly tend to be water-wet due to their hydrophilic nature.
    3.2 Relative Permeability
    3.3 Relative permeability of sand to flow of oil is higher than to flow of water.

3.2.2 Geological Deposition of Sand

Sandstones are the most important hydrocarbon reservoir, accounting for over 60% of the world's reserves in the giant fields. Excluding the Middle East, sandstone accounts for over 80% of the reserves. Sandstones are formed in a broad range of conditions and composed of relatively stable components less subject to physical and chemical changes than carbonate reservoirs. Below are listed the main depositional environments by which sandstones were originally formed.

3.2.2.1 Desert aeolian sands

The thickness may be large, similar to the modern Sahara desert, as much as 300 m thick. Usually, this type of sand is well sorted with well-rounded grains, with high porosity and permeability.

3.2.2.2 Marine shelf sand

Extremely wide range of reservoir thicknesses and lateral sizes due to constantly changing dispositional forces and direction of these forces. Variation of grain sizes and sorting causes wide variation in permeabilities and porosities.

3.2.2.3 Beaches, barriers and bar
These types of deposits are commonly 10–20 meters wide and 4 km long with bed thicknesses several meters. Grain sorting varies, becoming larder-grained and better sorted toward the top of the deposit.

3.2.2.4 Tidal flat and estuarine sands
These are typically deposited by reversing currents and are dominated by muddy currents. Thus low porosities and permeabilities make these types of sand poor producers.

3.2.2.5 Fluviatile sands
As a result of deposition by the river and their flood plains, these deposits vary greatly in thickness, width and depth such sands may typically range from 6–75m thick, 250m–20km wide and 6–160 km long with porosities 10–25% and permeability's from 60–2000 md.

3.2.2.6 Alluvial sands
These types of deposits are typically found at the bottom of the mountain chain where the runoff has been deposited at the base of mountains.

A deposit may be as much as 3000m thick varying in grain size from boulder catastrophically.

3.2.3 Formation Sand Description
A good concept of what the formation looks like and how it behaves down-hole is important. There is such a wide variety of formations and that correctly identifying the type of sand that is present will help determine how to deal with them. It may be categorized into three broad categories relative to its strengths.

3.2.3.1 Quicksand
Quicksand is a term often applied to completely unconsolidated formation sands. This type of sand has no effective cementing agent and is held together only by small cohesive force and compaction. It is difficult to drill through this type of formation as it readily collapses into the wellbore and tick the bit. Sand production begins immediately with fluid production if some means of

sand control is not used and flows readily with oil, water and gas. Quicksand can be found inland (on riverbanks, near lakes or in marshes) or near the coast. It can also form when an earthquake increases groundwater pressure, forcing the water to the surface and causing soil liquefaction.

3.2.3.2 Partially consolidated sand
The second type of formation sand is partially consolidated sand. It has some cementing agents but is only weakly consolidated. A core can usually be taken from this type of formation with a convention core barrel, but the core crumbles easily.

An open hole completion is possible with this type of sand, but without some means of sand control, the hole would collapse. Behind casing, this type of formation crumbles, forming small cavities of sand to come into the wellbore which readily fills the rat hole and forms bridges in the tubing.

3.2.3.3 Friable sand
The third type of potentially trouble-some sand is friable or semi-competent sand which is well cemented and easily cored. Cores of this sand appear strong enough and they don't look like they would allow sand to produce. However, produced fluid or gas readily erodes the sand from the face of the formation as it flows into the perforation. Open hole completion can be successful in this type of formation as the flow of fluid is spread out over a much larger area of the formation face thus the erosion forces are less.

3.3 Causes and Effects of Sand Production

3.3.1 Causes of Sand Production
The sand production mechanism is complex and influenced by each operation from the first-bit penetration of the producing zone to the start of production. The reason can be anyone or a combination of the following.

3.3.1.1 Totally unconsolidated formation
Some formations are totally unconsolidated or in a fluid state. Any attempt to produce formation fluid can result in production of a large amount of sand with the fluid.

3.3.1.2 High production rates
Some well produce sand if the production rate is high. These wells will produce sand production is restricting.

3.3.1.3 Water productions

In some formation, the cementation material is clay minerals and silt, which may be seriously affected by the water. When water production starts the bond is weakened or destroyed and sand may be produced.

3.3.1.4 Increase in water production

Generally, as water production increase, the total fluid production increased to maintain the maximum oil production .this increased production rate cause excessive stress on the weakly consolidated formation and may exceed the ability of the cementation material to bond sand together.

3.3.1.5 Reservoir depletion

In some cases, reservoir pressure is believed to add in the support of the overburden pressure. Reduction of reservoir or may cause the overburden to crush the formation and result in serious sand production.

3.3.2 Effects of Sand Production

The first problem with sand production is sand accumulation. A well—sands up when sand bridging occurs in the casing or tubing preventing flow. This happens when the sand load in the fluid becomes large and during shutdowns, it settles to the bottom. This sandy sludge may be hard to move and the perforations can gradually be covered over.

Sand also accumulates in downhole tools such as gas lift valves, sliding sleeves and sub-surface safety valves plus wellhead valves, causing operating failure. On the surface, sand accumulates in flowlines, separators and storage tanks. Separators have to be specially designed to cater to this sand problem.

Sand erosion is particularly a problem with chokes and flowlines. The problems of erosion are of greatest significance in gas wells where high velocities are encountered. The problems in oil wells are much less severe.

The casing can deform, buckle and fail due to sand production. This is unusual except in very long intervals of unconsolidated sand where there has been a long history of sand production and the casing takes a substantial loading from unconsolidated sand. Also, creation of void spaces behind the casing due to sand removal, can leave the casing unsupported and is subjected to buckling. This creates problems for subsequent workover operation and installation of downhole equipment.

3.4 Detection and Prediction of Sand Production

Several methods are used for predicting the likelihood of whether a sandstone reservoir will produce sand. Prediction methods vary in complexity and no method is completely reliable. All methods depend on various reservoir characteristics, such as drilling rate, density, modulus of elasticity, drill stem test data, logs, bulk compressibility and other characteristics. Conditions such as drawdown pressures, surge pressures, and treating fluid types may vary from well to well, limiting the reliability of these prediction methods. Nevertheless, the following information sources are suggested for predicting sand production: (1) experience in the area, (2) drilling data, (3) core-sample evaluation, (4) drill stem tests, (5) logs and (6) production data.

3.4.1 Methods for Monitoring and Detection of Sand Production

3.4.1.1 Wellhead shakeouts

In a wellhead shakeout a sample of produced fluid is placed into a graduated cylinder and centrifuged. Sand settles to the bottom and can be read as a percentage of the produced fluid. Although this is an accurate measurement of the sand in a given sample, it must be emphasized that this is a point sample that may not be representative of the actual overall production.

3.4.1.2 Safety plugs and erosion sand probes

These devices are thin-walled; metal flow line inserts that indicate pipeline erosion by cutting out before erosion critically weakens the flow line. Erosion probes are hollow steel cylinders of various lengths with one closed end which is inserted to the entrance of the flow stream. When the wall of the probe is penetrated by erosion, the flow stream pressure is transferred to a pilot valve which may sound an alarm or shut the well in. After a specified number of cutouts, the flow line is inspected ultrasonically to determine the actual extent of erosion. Safety plugs are thin-walled spots in the flow line to which a bull plug equipped with a pressure gauge is welded. When sand erosion wears through this thin section of pipe, the pressure gauge reads line pressure. Unlike erosion probes, safety plugs are not replaceable. Neither device can actually measure the sand producing rate but, instead, shows the effect of cumulative sand production.

3.4.1.3 Sonic sand detection

The acoustic sand probe, a development of Mobil Oil Company, detects the noise of sand impinging on the sensor and reports a signal proportional to the amount of sand for a given line size, fluid density, gas-oil ratio and sand size. This probe is very good at showing changes in sand production; but because of the many variables involved and its erratic behavior in multiphase systems, it is difficult to calibrate accurately for more than one well at a time. These probes are more expensive than erosion probes.

3.5 Methods for Sand Exclusion

Four broad groups of sand exclusion methods are available: (a) production restriction methods, (b) mechanical methods, (c) in-situ chemical consolidation methods and (d) combination methods. With the exception of production restriction methods, the remaining methods provide some means of mechanical support for the formation and help prevent formation movement during stresses that result from fluid flow or pressure drop in the reservoir.

3.5.1 Production Restriction

In sandstone formation, stresses resulting from fluid production or pressure drop act on the minerals that bind the sand grains together, resulting in sand production. One means of reducing this sand production is to restrict the production rate. This method has the lowest initial cost and in some cases, maybe a successful alternative to other available methods. However, in most cases, it is not a durable or economical solution to formation sand production.

Case histories of horizontal wells have verified that production rate restriction can initially reduce sand production. When a long interval of formation is exposed, equivalent flow volumes can be produced with much lower fluid velocities in the formation. Thus, controlling production in horizontal wells has provided a viable sand-control technique that can be used in formations that are completed with vertical wells.

However, in many instances, the reduced production rate that prevents the formation of sand is not profitable. In addition, the production rate is not always the only factor contributing to sand production. The degree of consolidation of the formation, the type and amount of cementitious material present, and the amount of water being produced are also significant contributing factors to sand production. These other factors may allow a well to produce sand even after the production rate has been severely restricted.

3.5.2 Mechanical Methods

Mechanical methods are the most common well treatments for excluding sand production. Mechanical sand-control methods are diverse, but they always include some type of device installed downhole that bridges or filters the sand out of the produced fluids or gases. These devices include a wide array of slotted liners, wire-wrapped screens and prepacked screens, which are generally used with gravel-pack procedures.

Screens and slotted liners encompass a broad range of downhole filtration devices. These devices (1) filter out the formation sand, (2) retain any particulate materials, graded sand, or other proppants placed against the formation to support it and (3) strain out the naturally loose component grains. Variations in screen designs, which will be discussed later, are affected by cost, durability and flow-through characteristics.

Gravel-packing is a mechanical bridging technique that involves placing and tightly packing a large volume of carefully graded proppant between the formation face or perforation tunnels. A filter device is used that is fine enough to retain the proppant in direct contact with the formation sand, thus preventing its movement toward the wellbore. The packed proppant is supported in the wellbore by a slotted liner or screening device. Advanced gravel-pack designs serve the dual purpose of stopping sand movement and providing production enhancement.

3.5.3 In-Situ Chemical Consolidation Methods

Sand control by chemical consolidation involves the process of injecting plastics or plastic-forming chemicals into the naturally unconsolidated formation, which provides grain-to-grain cementation. The objective of formation sand consolidation is to cement sand grains together at the contact points, maintaining maximum permeability.

3.5.4 Combination Methods

Methods in this category combine technologies of both chemical consolidation and mechanical sand-control; a sand filter bed is one example. In these processes, a gravel-pack is performed with a resin-coated pack-sand as the 'gravel'. Instead of being held in place with a screen or liner, the pack-sand is held in place by the cured resin, which provides a compressive strength of several hundred to several thousand pounds per square inch. Although procedures vary, the objective is to secure the pack in place while leaving the casing unobstructed. Combination methods include the use of products such as semi cured, resin-coated proppants and liquid, resin-coated gravel-pack sands.

3.5.5 Selecting the Appropriate Sand Exclusion Method

Once a well is identified as requiring sand exclusion, the appropriate exclusion method must be determined based on the following criteria:

1. **Economics**—the initial cost of the treatment and its effect on production
2. **Historical success**
3. **Applicability**—degree of difficulty to perform treatment
4. **Length of service**—estimations of sand-free production and need/frequency rates for repetition of the treatment.

Choosing the appropriate technique for sand exclusion requires an in-depth understanding of each sand exclusion method and its many modifications and variations. Table 3.1 shows the merits and limitations of the many types of mechanical and chemical consolidation methods covered in this chapter.

3.6 Mechanical Methods of Sand Exclusions

Gravel-packing techniques were developed for the water well industry in the early 1900s. The early gravel-pack techniques adopted by the petroleum industry consisted of running a slotted liner to depth, then pouring gravel down the annulus from the surface. This technique was sufficient on shallow, straight wells, but on deeper wells, the gravel would either not reach the bottom or it would bridge in the wellbore.

For improved gravel placement, specialized tools were developed. Tools such as set shoes, washdown shoes, clutch assemblies and stuffing boxes allowed the downhole placement of gravel through such techniques as the *washdown* or *reverse-circulation* methods.

As gravel placement problems persisted, tool systems were developed that allowed gravel to be pumped down the work string at greater velocities, which helps prevent premature bridging. The new tool systems also helped minimize contamination of the gravel with debris from the annulus, the premature bridging of sand in the annulus and stuck pipe. Tools developed for this system include the cup-packer crossover, circulating crossovers and hook wall packer systems.

3.6 Mechanical Methods of Sand Exclusions

Table 3.1 Merits and Limitations of Sand Consolidation Methods

Formation Characteristics	Mechanical Methods, Gravel-Packs, and Screens	Chemical and Combination Methods, Consolidation with Resins and Resin-Coated Sands
Formation Strength	Will not change formation strength	Adds considerable formation strength, with exception of theresin-coated sand
Permeability	Applicable. Certain techniques may reduce permeability	Applicable. Certain techniques may reduce permeability
Poorly Sorted Grain Sizing	Applicable using special job designs	Applicable with few restrictions
<10% Fines and Clays	Very Applicable. Good anticipated job life	Very Applicable. Good anticipated job life
>10% Fines and Clays	Applicable using special job techniques	Marginally applicable. Good resin injection and coverage is difficult
>10% Acid Solubility	Applicable with restricted acid pretreatments	Not applicable with acid-hardened type resins.
<10% Acid Solubility	Very Applicable. Good anticipated job life	Very Applicable. Good anticipated job life
<50° Hole Angle	Very Applicable. Good anticipated job life	Very Applicable. Good anticipated job life
>50° Hole Angle	Applicable using special tools, screens and techniques	Not applicable. Poor job success history, uniform coverage problems
Open Hole	Applicable using special job techniques	Not applicable. Poor job success history, uniform coverage problems
Cased Hole	Very Applicable. Good anticipated job life	Usually very applicable. Good anticipated job life
Slim Casing	Marginally applicable. Severe tool and screen restrictions	Very Applicable. Good anticipated job life
Single Zone	Very Applicable. Good anticipated job life	Very Applicable. Good anticipated job life
Multiple Zones	Applicable using special tools, screens and techniques, leaving screen in wellbore	Very Applicable. Leaves clear wellbore, should be done as an initial measure
<30 ft Interval Length	Very Applicable. Good anticipated job life	In most instances. Very Applicable. Good anticipated job life
>30 ft Interval Length	Very Applicable. May require special tools, screens, and designs	Not applicable. High costs and uneven resin coverage

Continued

Table 3.1 Continued

High Water Producer	Applicable. May require additional chemical fines control	Very Applicable. Good anticipated job life
Gas Producer	Very Applicable. Good anticipated job life	Applicable. Some resin systems clean up better with good anticipated job life
Oil Producer	Very Applicable. Job life depends on screen and proppant quality.	Very Applicable. Anticipated job life of 3 to 8 years
Low BHST, <120 °F	Very Applicable. Good anticipated job life	Provisional. Difficult curing conditions for some resins.
Medium BHST	Very Applicable. Good anticipated job life	Very Applicable. Good anticipated job life
High BHST, >250 °F	Provisional. May require special proppant and screen alloys	Provisional. Limited placement time for some resins; durability reduced
Steam Injection	Provisional. Likely will require special proppant and screen alloys	Marginally applicable. Some resins are more resistant than others

During the 1960s and 1970s, systems were developed in which the gravel-pack screen, production packer and circulating service tools were run in the well in one trip. Work string reciprocation was used to operate the mechanically set packer in the squeeze, circulating and reverse positions. When the gravel pack was complete, the crossover service tool assembly was retrieved and replaced by the production seal assembly.

During the 1980s, tools were further developed, and major advancements in carrier fluids were made. Filtered gelled polymers, such as hydroxyethylcellulose (HEC) were used for suspending the sand during pumping. Slurry sand concentrations could be increased to over 15 lb/gal, which both reduced pumping time and improved gravel placement into the formation tunnels for a more effective pack that did not contain commingled pack and formation sands.

In the 1990s, industry experts introduced the concept of increasing the pack-sand volume placed outside the casing (High-Permeability Fracturing). These *fracpacks* required even more specialized tool designs to withstand the high rates and volumes that were being pumped at high pressure. Synthetic proppants became more frequently used since they were more resistant to crushing and had higher permeability under high confining stress. However, because synthetic proppants are significantly more erosive than sands, they pose additional design problems for tool designers. Sand control continues to evolve. New gravel-pack systems, fluids and chemicals are continually being developed for improved sand placement and pack performance.

Four main components of a gravel-pack must be considered: pack- sand, screen type, carrier fluid, and tools.

3.6.1 Mechanical Components
3.6.1.1 Pack-sands
For a successful gravel-pack, the pack proppant must be carefully selected. The American Petroleum Institute (API) has issued document RP-58 (1995), which presents recommended practices concerning the evaluation of pack-sand. Accepted by most oil and service companies, this document sets quality standards regarding sand size gradations, shape (sphericity and roundness), strength and allowable percentages of foreign material in the sand that will be used for gravel-packing.

Sand size should be quality-controlled by sieving so that a proper formation sand-to-pack gravel size ratio is maintained and absolute permeability is not reduced. Grain roundness is a measure of the relative sharpness of grain corners, or of grain curvature (how nearly the edges of the grains approach that of a circle). Roundness is critical because angular sand has a greater tendency to form premature bridges than rounded sands. Angular sand is also more likely to chip and fragment during placement, which reduces the permeability of the pack and can clog slots and screens.

3.6.1.2 Liners and screens
Slotted liners and screens are downhole filters that provide different mechanisms and levels of sand retention or pack-sand support. Before a liner or screen is chosen, the well must be carefully evaluated so that the most applicable product can be selected. Screen construction and shape can influence (1) how well sand becomes packed in the annulus, (2) the flow capacity of the covered zone and (3) how long the composite pack might last.

The simplest slotted liner is made of oilfield pipe that has been slotted with a precision saw or mill. These slots must be cut circumferentially; otherwise, the tubing would become weak under tension. The individual slots can be as small as 0.020 in. or as large as required for the gravel size that will be used.

Mill-slotted pipe provides strength and economical service; it is particularly well-suited for water wells. Generally, slotted liners are used for oilfield applications only when wire-wrapped screens cannot be used economically. For example, slotted liners are typically used in wells that have long completion intervals or low productivity. Although slotted liners are much less expensive than wire-wrapped screens, they do not have high-inlet-flow areas and thus are not as useful for high-rate wells.

Wire-wrapped screens consist of keystone-shaped, stainless-steel or corrosion-resistant wire wrapped around a drilled or slotted mandrel made of standard oilfield tubular goods or special alloys. Spacing or standoff between the mandrel and wire wrap allows for maximum flow through the screen. In some screens, this spacing is created by grooves cut into the pipe; in other screens, ribs are affixed to the pipe (Figure 3.2).

In a prepacked screen, the annulus between the outer jacket and the pipe base is packed with gravel-pack sand. This sand may or may not be resin-coated. Three types of prepacked screens are available: a dual-wrapped prepack, a casing external prepack and a low-profile screen.

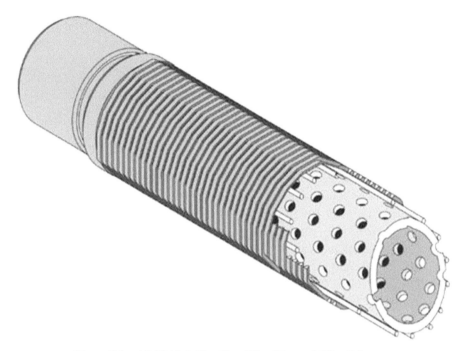

Figure 3.2 All-Welded, Pipe-Base Wire-Wrapped Oilwell Screen.

3.6 Mechanical Methods of Sand Exclusions

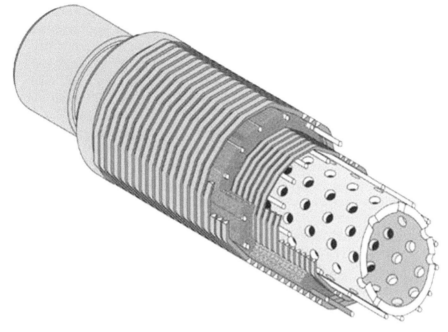

Figure 3.3 Casing-External Prepacked Gravel-Pack Screen.

The oldest screen type is a dual-wrapped pipe-based screen prepack. This screen provides built-in sand control when a gravel-pack fails or is not feasible (Figure 3.3).

The casing-external prepack screen has an inner wire-wrapped pipe base centered inside a large piece of perforated pipe. Primarily used for horizontal wells, this type of prepack screen is packed with a graded resin-coated pack-sand that is sized to bridge formation sand. This resin-coated pack-sand, cured at 350 °F, bonds the sand grains together to prevent the sand from coming out the perforations of the outer perforated pipe.

The third type of prepack screen is the low-profile screen. This screen is similar to the dual-screen prepack, but it consists of an inner micro screen and a regular outer screen jacket

The pre packed sand layer is very thin compared to regular prepacked screens. Low-profile screens are recommended for wells that will be gravel-packed but in which casing sizes and tubing requirements restrict the outside diameter of the screen. The thin layer of gravel between the jacket and the pipe base helps ensure against possible voids in the gravel pack.

Alternative screen designs are continually being developed that provide more durability, compensate for voids left in the gravel pack or allow better sand-packing while preventing premature standouts.

The sintered metal screen was designed as an improvement over low-profile prepacks. Instead of the prepacked sand layer around the center pipe, a jacket of seamless, porous, sintered stainless steel is used. The uniform porosity of the metal allows the use of any grade of gravel-pack sand.

The woven metal-wrapped screen was introduced to fill a similar need as the sintered metal screen, and it can also be used with any grade of gravel-pack sand. This screen consists of layers of various grades of woven metal that are wrapped around a perforated center pipe, which allows the packed zone to be post-acidized without damaging any pre packed sand layer in the screen.

The auger-head screen design has been used as a conventional screen with a wash down shoe. The auger end allows the screen to be screwed into the sand-filled casing or hole. As the hole angle becomes more horizontal, this design becomes more effective than conventional wash down configurations.

Another alternative screen design, the shunt tool, reduces premature standouts during the packing of very long intervals. This screen design provides an extra shunt path for gel-sand slurry, which reduces the possibility of slurry dehydration when high-permeability streaks are packed over the composite interval.

3.6.1.3 Carrier fluids

A fluid used for gravel-packing has three primary functions. First, it must transport the sand or pack medium to the location in the well where the gravel pack will be established. During a horizontal gravel-pack treatment, the packing grains will change direction several times. The velocity or viscosity of the fluid must be able to influence the direction of the grains. Without a transport medium, the grains will always travel downward. If sufficient lift is not provided throughout the placement, premature sandout could occur. The Sandout is caused when an agglomeration of grains develops in critical pathways, blocking the progress of the remaining grains necessary to form the desired pack.

The second function of carrier fluids is to separate themselves from the grains to allow the close contact desired. The degree to which this fluid loss occurs is critical to gravel-packing success: if the fluid separates too early or too much, a premature sandout may occur; if the fluid separates too late or incompletely, a void may be left when the pack grains settle out. As the grains arrive at the packing location in the well, they must be deposited compactly and sequentially against the formation. The fluid may either exit through the screen and wash pipe and return to the surface or, more conventionally, exit to the adjacent formation.

The important third function of a gravel-pack fluid is to return from the formation without reducing permeability within the zone treated. When the gravel pack is finally established, the operator will want to put the well on production. The gravel-pack fluid lost to the formation must now change direction and flow back into the wellbore. Often, a fluid property that is desirable for grain transport may prevent the fluid from rapidly returning from the formation. A high-viscosity fluid is such an example. To rid a fluid of its viscosity before recovery from the formation, a gel breaker must be used. Once in the formation, the fluids have the opportunity to react with the formation grains. These reactions can increase the difficulty of recovering the packing fluid as well as attaining suitable hydrocarbon flow from the packed zone. For example, clay swelling can be caused when clays are exposed to low-salinity water. If the fluid contains surfactants, their influence on the water and hydrocarbon flow from the formation must also be considered. Certain packing fluids that include polymers to create viscosity will leave a chemical residue in the formation when the gel is broken by chemical breakers. The amount of gel residue associated with the different polymers varies greatly. Before a job is performed, the proposed polymer and mixing brine should be carefully analyzed. When the connate water in some oil-producing formations contacts the dense brines (zinc bromide) used in the gravel pack, a pH change could occur; specifically, zinc hydroxide could form, plugging the pores of the formation.

In addition to various brines that have been successfully used in gravel-packing, several polymers have been added to these brines to enhance the degree of sand transport and packing efficiency in gravel-packing operations. Some of the common polymers and their properties are listed in Table 3.2.

Table 3.2 Common Polymers Used in Gravel-Packing

Gelling Agent	Chemical Category	Common Trade Names
HEC	Hydroxyethyl cellulose	HEC-10, HYDROPAC, AQUAPAC
SGC	Succinoglycan, biopolymer	ShellFlo S, FLO-PAC
Xanthan	Xanthan, clarified biopolymer	Xanvis, Bi-O-PAC
Surfactant Gel	Ammonium quat	PermPAC AV

Gelled fluids made from the polymers in Table 3.3 and ungelled brines each have their champions in the industry.

Table 3.3 Gelled Fluids and Their Characteristics

Polymer	HEC	SGC	Xanthan	Surfactant Gel
Usage Level	60–80 lbm/gal	20–50 gal/gal	15–35 lbm/gal	20–40 gal/gal
Cost Factor[a]	1.00	1.07	1.50	2.52
Physical Form	Dry powder	Jelly-like liquid[b]	Dry powder	Pourable liquid[c]
Preparation Steps	Adjust to pH = 3–4, add polymer and raise pH to 7	Add over-the-top into high shear area; No pH adjustments.	Use Fe sequestering agent; Add polymer; add salts last.	Batch or continuous mix of surfactant with chloridebrines
Requirements for Optimum Performance	Preshearing and filtration to 5–10 μm	Complete dispersion of the polymer	Extensive and specific preshearing then filtration	Complete dispersion of surfactant
Restrictions	Does not yield in three-salt brines, mostly 15–18 lbm/gal	Slow to yield in brines with limited free water	Does not tolerate divalent brines, i.e. $CaCl_2$	Does not tolerate saturated brines or ones containing $ZnBr_2$
Maximum Viscosity at 75 °F (cp)	160	23	21	240
Sand Settling	Slight	Almost none	Almost none	Almost none
Minimum Viscosity at 75 °F (cp)	45	14–16	14–16	66
Sand Settling	Moderate	Slight to Moderate	Slight to Moderate	Almost none
Temperature Limit	190–230 °F, depending on the mix brine	160 °F–210 °F, Very dependent on mix brine.	190 °F–230 °F, Depending on the mix brine	200 °F
Common Classes of Breakers Used	Acids Oxidizers Specific Enzymes	Oxidizers Oxidizers with Enhancers	Oxidizers Oxidizers with Enhancers	No internal breakers. Broken upon dilution by formation fluids or flushes

Since good, long-lasting gravel-pack jobs have been reported with each of the contending fluids, it appears that technique is as important as fluid properties in the outcome of the gravel pack. Most research now shows that the best perforation packing can be achieved with a viscosified fluid. However, the same study shows that the best annulus packing is attained with ungelled brine. Staged gravel-pack treatments allow the use of both packing methods. First, a number of gelled brine-sand slurry stages are pumped to fill the perforations and then the treatment is completed with brine and low sand concentrations that pack the annulus.

3.6.2 Tools and Accessories

The tools and accessories for gravel-pack treatments either (1) remain in the well or (2) are run, placed and retrieved before the production tubing is run.

3.6.3 Completion Tools

3.6.3.1 Gravel-pack Packer
The gravel-pack packer is one of several packer types and designs. Generally, a sealbore-retrievable or permanent packer is used. If a retrievable packer contains slips above the element package, sand could settle on top of the packer, causing retrieval problems.

3.6.3.2 Flow sub
A flow sub can either be a ported sub or a sleeve-closure device combined with a sealbore and lower pup joint. The flow sub provides a location for the service-tool gravel exit ports and a path that directs flow to the outside of the screen. Closing sleeves are primarily used (1) if sufficient lengths of blank are not available to restrict flow up the annulus instead of through the screen and (2) if production seals cannot be run below the ported sub to prevent sand production.

3.6.3.3 Mechanical fluid-loss device
A mechanical fluid-loss device stops uncontrolled fluid losses to the formation while the service tools are being tripped out of the well and production tubing is being run. These tools can be any one of a number of the flappers, ball, or plug-type devices that can later be broken, dissolved or expended.

3.6.3.4 Safety joint

A safety joint allows operators to detach the screen from the packer during retrieval operations. It also provides a safeguard against extreme loads being placed on the gravel-pack packer as formation compaction or gravel-pack settling occurs. This tool should maintain the integrity between the inner and outer mandrels until approximately 12 in. of movement occurs.

3.6.3.5 Blank pipe

Blank pipe above the screen provides a reservoir of proppant that can fill in from the top of the screen as settling occurs. Generally, at least 60 ft (two joints of pipe) should be used on high-rate water-packs (HRWPs) and fracpacks. Gravel packs using high-density slurry should include enough blank pipe above the screen to accommodate the entire slurry volume in the annular space below the crossover. If less than 60 ft of blank pipe is used, the following issues must be considered:

1. Gravel-pack quality (compacted, void-free pack)
2. Screen damage from high sandout pressures or direct impingement from adjacent perforations
3. Carrier fluid viscosity (sandout can proceed up to and into the work string)

Production screens are installed next in the string.

3.6.3.6 Tell-tale screen

A tell-tale screen is a short, upper or lower section of the screen that indicates when the proppant pack has achieved a certain height during pumping. It is separated from the rest of the screen assembly by a seal that opens or closes, circulating flow as the gravel-pack service tool is manipulated. In new applications, tell-tale screens have been virtually replaced by large-diameter wash pipe.

3.6.3.7 Seal assembly

The seal assembly provides a seal between the bottom of the screen assembly and the sump packer. It is typically run with a slotted cylindrical collet, which has a raised diameter that interferes with the inside of the sump packer sealbore. This collet indicates the presence of drag when the seal assembly passes through the sealbore of the sump packer.

3.6.3.8 Sump packer

The sump packer is the lowermost packer in the sand-control completion assembly. A variety of packer designs can function as a sump packer, which must provide a solid bottom that can contain the proppant at the lower end of the gravel-pack interval. Without this packer, any proppant settling in the rat hole could result in voids in the proppant pack around the screen. Packers normally used in this application are permanent or retrievable sealbore packers (Completion Hardware) that are run on electric line or hydraulic setting tools, generally before perforating. This packer also provides a location where weight can be applied to position the gravel-pack assembly in the wellbore.

3.6.4 Service Tools

3.6.4.1 Crossover service tool

The crossover service tool provides channels for the circulation of proppant slurry to the outside of the screen and returns circulation of fluid through the screen and up the wash pipe. Most tools also include the hydraulic setting tool that sets the gravel-pack packer.

3.6.4.2 Reverse-ball check-valve

A reverse-ball check-valve can be as simple as a ball sitting on a restricted-diameter crossover in the service tool or as complex as a multiple-actuating ball valve. Regardless of the valve design used, the valve should not create a hydraulic lock or result in excessive differentials to the formation.

3.6.4.3 Swivel joint

The swivel joint allows operators to assemble concentric tubing strings by freely rotating the joint as threads are made upon the outer strings of the pipe.

3.6.4.4 Washpipe

The wash pipe is attached to the gravel-pack service tool and run inside the screen. The wash pipe serves two functions. First, it provides a return fluid circulation path that can be spaced out at the very end of the screening interval. This path forces the proppant slurry to flow to the lowermost screen before bridging in the screen-casing annulus.

The second function of the wash pipe is to prevent the proppant carrier fluid from flowing to the outside of the screen. Loss of fluid from proppant slurry can cause premature and rapid bridging to occur, especially in high-density proppant concentrations. A number of studies have verified that the pipe diameter should be at least 80% of the inside diameter of the screen base pipe.

3.6.4.5 Shifting tools
Shifting tools position the closing sleeves or close the flapper valves as the service tool is pulled from the well.

3.6.4.6 Tool selection
once the well objectives are understood, a completion technique and well completion tools can be selected. The service provider and the tool supplier must know the casing size and weight, zone depths, bottom hole pressure, expected pressure differential and the well fluids (both completion fluids and produced fluids). Generally, a packer is selected first on the basis of the size, weight, and grade of the casing. The bore size of the packer is typically influenced by the tubing size used, which affects the size of many accessory items. Because well fluids are occasionally corrosive, the packer and completion tools must be corrosion-resistant.

3.7 Mechanical Method: Techniques and Procedures

A complete, void-free gravel pack is one of the most effective sand-control measures available. The annular portion of the pack alone cannot sustain high-rate well productivity over a long period. The external gravel pack (the area either in a perforation tunnel or fracture that extends past any near-wellbore damage) is a key to prolonged trouble-free production. Since the perforation tunnels are the only communication from the formation to the wellbore, they must remain unclogged by formation sand and fines. If gravel-packing is used for controlling sand production, the pack-sand should fill the perforation tunnels and void spaces behind the casing. Sometimes, the formation can even be restressed, although restressing is seldom achieved except through pressure gravel-packing.

1. Gravity Pack
2. Washdown Method
3. Circulation Packs
4. Reverse-Circulation Pack
5. Bullhead Pressure Packs
6. Circulating-Pressure Packs
7. Slurry Packs

8. Staged Prepacks and Acid Prepacks
  9. Water-Packs and High-Rate Water-Packs
 10. Fracpacks

3.7.1 Gravity Pack

A gravity pack is one of the most primitive gravel-packing techniques. With water in the hole, a screening device is run into the hole on tubing with a backoff joint. Sand is then slowly poured down the annulus while fluid is running into the hole. This technique is one of the least expensive approaches, and it provides very little control of sand placement. In fact, it is unlikely that any sand is forced through the perforations; sand may bridge around the tubing collars during sand placement. Consequently, when this method is used, flow into the wellbore may quickly become restricted by formation fines invading the pack.

3.7.2 Washdown Method

The washdown method consists of depositing gravel to a predetermined height above perforations, then running the screen and liner assembly with a wash pipe and a circulating shoe. The screen is washed down through the gravel. When the shoe is on the bottom, gravel is allowed to settle back around the screen and liner. The most basic washdown tool system has minimal provisions for compacting the gravel in the screen-casing annulus and no means of squeezing gravel through the perforations. However, perforations are often pressure-packed before the screen is washed down. In more complex tool systems, the tools can perform the washdown technique followed by a circulating, squeeze or fracpack after the packer has been set.

During the washdown procedure, the gravel sizes tend to segregate. In addition, the tools could stick or stop short of the bottom. If viscous fluids are used as a means of suspending the sand, they may damage the zone.

3.7.3 Circulation Packs

Circulation packing, sometimes called conventional gravel-packing, normally involves the placement of gravel that is suspended in a low-viscosity transport fluid pumped at low gravel concentrations. Circulation packing is usually conducted in an open hole or in a cased hole after the perforations have been prepacked with gravel. Typically, the transport fluid is filtered brine with gravel added at a concentration of 0.5–1.0 lb/gal. The gravel is commonly mixed into the fluid through a gravel injector while fluid is pumped at approximately 0.5–3.0 bbl/min. Gravel is transported into the annulus between the screen and casing (or the screen and the open hole), where it is packed into position from the bottom of the completion interval upward. The transport fluid then returns to the annulus through the washpipe inside the screen that is connected to the workstring. Circulation packing is compatible with essentially all the subsequent placement techniques.

3.7.4 Reverse-circulation Pack

As the term implies, the *reverse-circulation method* involves the reverse circulation of water-sand slurry or gelled water-sand slurry.

For the placement of some sand through the perforations, returns can be shut off and pumping can be continued until a pressure increase indicates that sand has covered the tell-tale screen. If no positive pressure is exerted, only sand-packing inside the wellbore is likely to be achieved. Two or three joints of unslotted tubing are generally run above the screen. These tubing joints provide a space for sand to fill the annulus above the screen, which normally prevents fluid and sand movement up the annulus and provides a reserve of pack-sand if the pack should settle.

3.7.5 Bullhead Pressure Packs

This inexpensive procedure requires a packer and a releasing crossover tool; a tell-tale screen or wash pipe is not required, since no returns are taken during the procedure. The water-gel sand slurry is pumped down the tubing behind a gel prepared. Slurry is then directed from the tubing into the casing-screen annulus through the releasing crossover tool below the packer. Pumping pressure increases as more perforations are covered with pack-sand. After the final sandout, the packer is released from the screen and blank pipe at the crossover tool.

3.7.6 Circulating-pressure Packs

A *circulating pressure pack* requires a packer and crossover tool as well as a screen with an internal washpipe. This technique allows slurry to be injected down the tubing, which prevents the scouring of drilling fluid, rust, pipe dope and scale from the tubing-casing annulus and reduces the potential of pack permeability damage. The slurry crosses over into the annulus below the packer. The carrier fluid deposits the sand and enters the screen and wash pipe. Fluid is then conveyed to the annulus above the packer by means of the crossover and then returned to the surface.

For sand to be placed through the perforations, the surface returns should be stopped at some point and the slurry should be squeezed against the formation before the perforations are covered with slurry.

3.7.7 Slurry Packs

Slurry packing can be performed with a variety of tools and pumping techniques. The procedure consists of pumping gravel at high concentrations in a viscous transport fluid. The slurry is usually batch-mixed in a blender or paddle tank before it is pumped. Although the fluid can be either water- or oil-based, HEC viscosities brine is the usual choice. The typical polymer loading is 60–80 lb/Mgal. In some situations, the gelled fluid is cross linked so that it creates very high viscosities. The concentrations of gravel usually pumped with slurry-pack fluids is about 10 lb/gal. In certain situations, however, concentrations have ranged from much less than 10 to more than 18 lb/gal. When these high-density fluids are pumped, the fluid and gravel tend to move as a mass. Compared with the low-density conventional gravel-pack fluid, these slurry systems have significantly greater gravel-suspending capabilities. Depending on well conditions, pump rates normally range from 1/2–4 or 5 bbl/min when gravel is placed around the screen.

When high-viscosity fluids transport gravel into the completion interval, a reserve volume must be specified. To allow for subsequent resettling, this volume is usually higher than the reserve volume required for low-viscosity fluids. Well conditions affect the gravel reserve volume when viscous fluids are used. Before the gravel is pumped, the service provider should calculate the theoretical amount of gravel required to pack the annulus around the screen and the amount of gravel reserve. Placement should be approximately 100% under ideal conditions. Depending on the additional gravel placed through the perforations, possible bridging in the tubular or pumping conditions, the amount of gravel placed could differ from the theoretical volume. If the amount of gravel is considerably less, premature bridging has probably occurred. Therefore, before the well is placed on production, gravel must be settled around the screen and additional gravel must be pumped.

When high-viscosity fluids circulate gravel around the screen, the pump rate should be low enough that the viscous drag forces do not exceed the gravitational forces once the slurry has reached the completion interval. If the viscous forces become dominant, the gravel will be drawn into the screen rather than settling to the bottom of the well. As a result, the gravel will pack radially outward from the screen, possibly resulting in premature indications of a completed gravel pack. This type of packing geometry and sequence is not desirable; ideally, the gravel should dehydrate from the bottom of the completion interval upward. The potential for non-uniform packing is one of the disadvantages of slurry packs. Therefore, the actual volume of gravel pumped should be carefully compared to the theoretical volume. If the pumped-gravel volume is much less than the theoretical volume when high-viscosity fluids are used, the gravel should be allowed to settle before additional gravel is pumped for completing the pack.

An upper tell-tale screen should be avoided when high-viscosity fluids are used for circulating gravel because gravel will probably screen off on the upper tell-tale. This screen off could prevent gravel placement over the main completion interval. When no upper tell-tale is used, the screen or slotted liner should be extended above the completion interval before the blank tubing is added so that a gravel reserve is available. Because of gravel settling, part of the gravel reserve volume will be filled if well deviations do not exceed approximately 60°. Sometimes, a lower tell-tale screen is used with slurry packing. If a lower or upper tell-tale screen is not used with this system, the washpipe is attached to the bottom of the screen section.

3.7.8 Staged Prepacks and Acid Prepacks

Productivity can usually be increased when perforations are prepacked with gravel immediately before the major gravel pack is pumped. The prepack provides a more complete gravel fill of the perforation and as a result, it increases perforation permeability. The gravel pack prevents perforations from collapsing or filling with formation sand.

Best results were achieved when alternating stages of acid treatment and gravel slurry were displaced in volumes sufficient to treat 20–30 ft of the well interval. Therefore, if the treatment interval is 100 ft, then four or five phases of acid and slurry will be needed to provide enough gravel-pack material for the entire interval.

3.7.9 Water-packs and High-rate Water-packs

As its name suggests, a *water-pack* uses water as a carrier fluid for gravel-pack sand. Because the polymer residue from slurry packing could potentially damage formation permeability, water-packs have become a popular alternative to slurry packing in recent years. Water-packs can help form very tightly packed annular packs, but they have a high leak-off rate in high-permeability zones, which can result in bridging in the screen/casing annulus. This bridging can then cause premature screen out of the treatment.

High-rate water-packs were developed to overcome the high leak-off problems encountered with standard water-packs in high-permeability formations. High-rate water-packs are usually preceded by an acid prepack, and they place far more sand (up to 700 lb/ft of perforations) since they exceed the formation-parting pressure. Other successful high-rate water-pack treatments have been reported from geopressured reservoirs where very little differential has existed between static formation pressure and formation parting pressure.

Although the term *water-pack* suggests that only water is used for proppant transport, lightly gelled slurry (25 lb HEC/Mgal) is frequently pumped. The danger that zonal isolation may be compromised is possible with high-rate water-packs, just as it is with fracpack completion services. Tracer log data from some water-packed wells indicate that incomplete entry of the tracer over the entire interval height has occurred, leaving only the high-permeability area of the zone packed. Therefore, only wells with formations that can sufficiently resist fracture height growth are candidates for high-rate water-packing. Since the pumping rate, fluid volume, and pressure in this type of completion are large, the tools used must withstand the more severe well conditions. If tools are not properly selected, downhole equipment failures could be caused by the collapse and flow erosion. Both the fines in the fluid stream and the return flow rate must be controlled so that the possibility of screen erosion during the packing operation is limited.

3.7.10 Fracpacks

Fracpack treatments are a special type of gravel-packing in which the volume of sand placed outside the perforations far exceeds the volume of the voids behind the pipe. The theory and methods used to optimize this type of gravel-pack treatment are discussed in High- Permeability Fracturing.

3.7.11 Summary

A given technique is chosen on the basis of economics. A number of factors have a relationship to completion cost. These factors include

1. Design simplicity
2. Minimal rig time through the elimination of trips

3. Minimal rig floor assembly time
4. Well depth
5. Well control problems (fluid loss, high-pressure zones)
6. Zone spacing and the number of zones to be completed
7. Bottom hole pressure, temperature and fluids
8. Availability of surface pumping and blending equipment
9. Required well completion life
10. Work over costs
11. Safety considerations

Each of these items contributes to the system selection process. These decisions also vary geographically.

3.7.12 Mechanical Job Designs

Effective mechanical sand exclusion requires good job execution and a good gravel-pack design. The primary objective of the application should be to control the production of formation sand without excessively reducing well productivity. During the job design process, the parties involved must choose a pack-sand grade, a screening device, a carrier fluid, chemical pretreatments and placement techniques. The first step of the job design is to evaluate the formation that requires the gravel pack.

3.7.12.1 Formation characteristics

Formation permeability influences the type of carrier fluid that will optimize leak-off during pack placement. Sieve analysis can reveal the formation's average structural grain size, allowing job designers to determine the correct grain size of pack sand. Knowledge of the formation mineralogy, usually measured by X-ray diffraction instrumentation, helps job designers identify troublesome feldspars and clay minerals that are prone to swell and/or migrate as fluids are produced through the pack.

The relationship between the particle size distribution of given formation sand and the critical size required for gravel-pack sand is significant. Therefore, job designers must distinguish the size of the load-bearing solids from the size of formation fines. Fines are very small particles of loose solid materials in the pore spaces of nearly all sandstone reservoirs. Produced fines likely originate at the interface between the gravel pack and the near-well formation, rather than from distant points in the reservoir. The higher flow velocities near the well probably contribute to increased fines mobility in this region.

The integrity of the formation analysis depends on the quality of the sample used in the analysis. The most accurate and preferred sample type is a full core obtained during drilling across the expected interval that will be completed. However, this type of sample is expensive to obtain, especially if the exact completion interval is unknown, and long intervals of formation must be cored.

Bailed or produced samples are generally unacceptable because sample segregation can occur, resulting in produced samples that include finer material and bailed samples composed of the coarser fractions. These samples tend to represent an average of the particle sizes across the interval, and the representative average is usually larger than the finest sands in the formation.

Obtaining samples of the formation with a wireline sidewall coring tool or gun is the most popular method. The gun is run into the open hole before it is cased and a hollow core barrel is fired through the filter cake into the formation of interest. Therefore, the formation material obtained is of the immediate wellbore region. This formation material is usually flushed with drilling fluid filtrate that penetrates the formation before filter-cake buildup; therefore, cores that are taken subsequently are usually contaminated with drilling fluid clays. Contamination with these fines can lead to erroneous results during sample analysis. The amount of artificially introduced fines can be as high as 30% by weight. Since most of this material is 20m or smaller, sieve analysis of these samples will be skewed toward smaller size and result in a D_{50f} value that is much smaller than the true value; consequently, a finer pack-sand is recommended and lower productivity may result.

Two acceptable methods are used for determining the particle size of the formation sand. The first method is sieve analysis, in which a sample is cleaned of oils and information constituents, and then dried and passed through a specific set of sieves. For all comparative analyses, the industry has adopted the *US mesh series*, which consists of a standard series of 12 sieves and a bottom pan. In this series, each sieve opening has twice the cross-sectional area as the sieve below it in the series. Table 3.4 lists the sieve numbers and their opening sizes.

The sieve analysis is performed according to ASTM Procedure C135–84A, which requires that the portion of the sample retained on each sieve is weighed and that a weight percentage is calculated. When graphically displayed, the sieve analysis data can reflect the sizes of the component grains and their comparative contribution as a percentage of weight. Figure 3.4 shows how the D_{50f} point of an analysis is determined graphically.

100 Sand Control

Table 3.4 Standard Series of 12 Sieves for a Sample Analysis

US Sieve Number	Sieve Opening (in.)
10	0.0787
20	0.0331
30	0.0232
40	0.0165
60	0.0098
80	0.0070
100	0.0059
120	0.0049
140	0.0041
170	0.0035
200	0.0029
325	0.0017

This point represents the median grain diameter of the formation tested. Although other significant points along the sieve curve have been used for calculating a pack-sand size, especially where non-uniform sands are involved, the D_{50f} point is nearly universally accepted for this purpose.

A second method is to use an electronic particle counter. Suitable instruments for this task use either the principle of light blockage or laser beam technology to sort the constituent grains of the sample according to size. Some of these instruments have variable size ranges that can be examined. Most instrumentation uses selected size channels (usually designated in μ m) that roughly correspond to the sizes of the US mesh series of sieves discussed earlier. These instruments can calculate the D_{50} point and other points of the sample, and they can also produce the typical S-shaped curve shown in Figure 3.4, from which the D_{50f} point and grain size distribution of the sample can be observed. Instrumental sizing methods generally require smaller sample volumes, reduce the likelihood of human error and allow for more complete particle separation than measurements taken through the use of dry sieving (Figure 3.4).

3.7.12.2 Pack-sand selection criteria

One of the most important parameters in a gravel-pack design is the ratio of the gravel grain size, D_{50p}, to the formation sand grain size, D_{50f}. When the D_{50p}/D_{50f} ratio is high, the oversized pack-sand will allow the invasion of formation sand, which reduces the overall permeability of the packed zone (often to less than the native reservoir's permeability). Conversely, if undersized gravel is used, it will provide excellent sand control, but it may jeopardize productivity in certain situations.

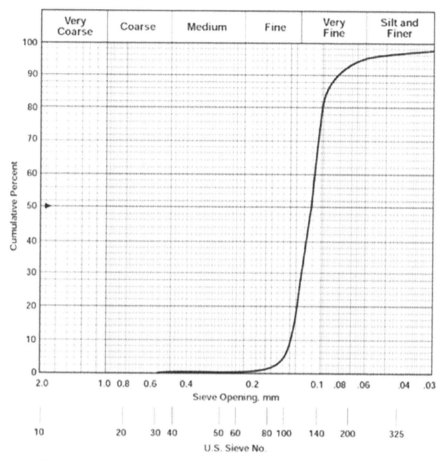

Figure 3.4 A Typical Sieve Analysis Plot Showing the Value of the D_{50f} Point (in.), Found by Extending a Perpendicular Line Down from the D_{50f} Point on the Curve.

Figure 3.5 shows that maximum permeability occurs when the D_{50p}/D_{50f} ratios are less than or greater than 10.

Specifically, a ratio of approximately 6 provides the maximum gravel-pack permeability with good sand control. At a ratio of 15, the pack permeability is good, but sand control is poor because the formation sand tends to move into the pack-sand. At a ratio of 10, the formation sand can move into the gravel pack, but it will have difficulty moving through it, causing a severe loss in overall productivity.

102 *Sand Control*

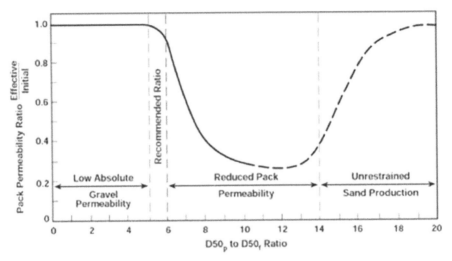

Figure 3.5 Effect of D50p/D50f Ratios on Sand Control and Pack Permeability, after Saucier (1974).

Many years of field-proven experience have shown that a D_{50p}/D_{50f} ratio of 5–6 helps compensate for possible sampling errors or a lack of samples from the entire zone in question. A suggested procedure for properly determining a workable grade of pack-sand is as follows (Figure 3.6):

1. Determine the D_{50f} value from a representative formation sample.
2. Multiply that value by 5, which results in the D_{50p}.
3. Compare the calculated D_{50p} required to the values in Table 3.5 (for API grades of pack-sand).
4. Select the grade of API-approved pack-sand from Table 5 that is nearest to the calculated value.

3.7.12.3 Screen selection criteria

The screen is not the primary source of sand stoppage in a gravel-pack completion. Instead, the gravel-pack sand is primarily responsible for stopping the formation sand from moving toward the wellbore. Therefore, the screen must be 100% effective in retaining the gravel-pack sand in place.

Screen quality covers a variety of properties: flow capacity, tensile strength, collapse strength and corrosion resistance (Figure 3.7).

Most high-quality oilwell screens are constructed of alloys that have maximum strength and resistance to the fluids in the well. Suggested alloys

3.7 Mechanical Method: Techniques and Procedures   103

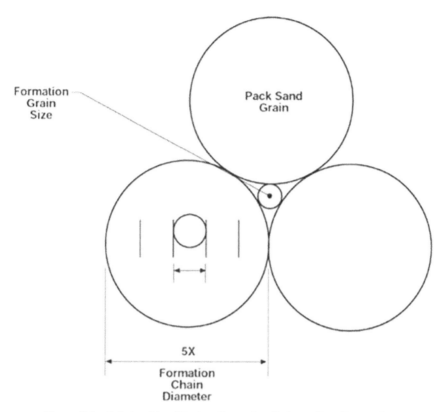

Figure 3.6 Relationship of Pack to Formation Grains at $D_{50p}/D_{50f} = 5$.

Table 3.5 Recommendations for Pack-Sands and Screen Devices

D_{50f} Times Factor 5–5.5 (in.)	Recommended Grade of Pack-Sand	Recommended Slot Width (in.)	Recommended Screen Wire Spacing (in.)
0–0.0125	50–70-mesh	NA	0.006
0.0125–0.017	40–60-mesh	NA	0.008
0.017–0.023	30–50-mesh	NA	0.01
0.023–0.030	20–40-mesh	NA	0.012
0.030–0.0455	16–30-mesh	0.016	0.016
0.0455–0.0595	12–20-mesh	0.025	0.025
0.0595–0.0715	10–12-mesh	0.035	0.035
0.0715 and Larger	8–12-mesh	0.05	0.05

104 *Sand Control*

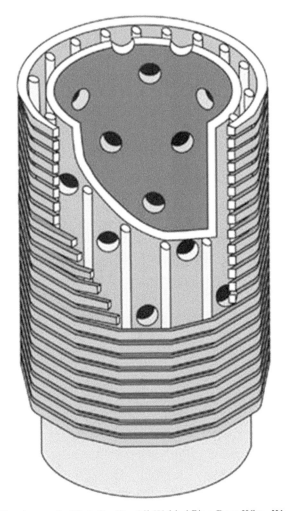

Figure 3.7 Cut-Away of a High Quality All-Welded Pipe-Base Wire-Wrapped Screen.

include 304 stainless steel, 316D stainless steel, and Incoloy 825 for high-temperature use. These screens include a densely perforated center pipe for support against collapse and gauge distortion. The keystone-shaped screen wire minimizes plugging from particles that are small enough to enter from the exterior. This keystone-shaped wire is wrapped with a smaller dimension toward the mandrel, which reduces clogging since only a minimal contact area exists between the wire and particle. The screen is joined by a high-strength weld with an external wrap wire to each vertical support rod. The gauge of the screen should be consistent throughout the tool, ranging from ± 0.001/0.002 in.

In addition to the above features, prepack screen quality can be assessed on the basis of the uniformity of its sand-pack bed; the bed should have no voids. Any movement of the prepack sand bed within the screen indicates a poorly packed product.

The length of the production screen depends on the length of the perforated interval. The general rule in high-density slurry packs is to use a 5ft overlap below the bottom of the perforated interval and a 5–12ft overlap above the top of the perforated interval. This overlap maximizes productivity and possibly compensates for any depth measurement errors involving the location of the sump packer in relation to the perforations. If the well deviation is increased to more than 40°–50° from vertical, the overlap above the top of the perforated interval should be increased. Experience in a particular field or formation will also influence the amount of top overlap required.

For proper gravel-pack installation and operation, an annular clearance of 3/4 in.–1 in. between the OD of the screen and the ID of the casing is necessary. The greater the annular clearance, the greater the pack placement and function efficiency. Pack placement problems and premature sand bridging in the annulus most frequently occur when the annular clearance is less than 3/4 in. Table 3.6 shows suggested screen OD measurements in relation to various casing sizes.

The annular clearance should allow the screen to be washed over easily during any subsequent workovers that might be required. Anything less than a 3/4-in. clearance is not recommended because of the extreme difficulty in washing over the screen.

Manufacturers of wire-wrapped screens express the space between the wires in units of 0.001 in., which is referred to as the *gauge* of the screen. The correct screen gauge is chosen according to the grade of pack-sand that the screen will have to retain. Since 100% retention of all of the pack-sand is essential during all phases of well life, the decision cannot be safely based on simple sand bridging.

The smallest pack-sand grains are represented by the highest mesh number in the grade designation. For example, 60-mesh (0.0097-in. diameter) is the smallest grain size in a 60/40 grade of pack-sand. To safely retain 60/40-mesh pack-sand, the screen gauge should be 0.5–0.9 times 0.0097 in. or 0.0049 in. to 0.0088 in. Therefore, a 6-gauge screen should be selected. The spacing between the wire should be 0.5–0.9 times the diameter of the smallest pack-sand grains (Figure 3.8).

Table 3.6 Recommended Screen Diameters for Adequate Gravel-Pack Annulus

Casing OD (in.)	4–1/2		5		5–1/2	
	Max. ID	Min. ID	Max. ID	Min. ID	Max. ID	Min. ID
	4.09	3.826	4.56	4.00	5.044	4.548
Base pipe OD (in.)	1.66	1.315	1.9	1.66	2.375	1.90
Screen Wire-Wrap OD (in.)	2.26	1.94	2.55	2.26	2.97	2.55
Clearance (in.)	0.92	0.94	1.01	0.87	1.04	1.00
Casing OD (in.)	6		6–5/8		7	
	Max. ID	Min. ID	Max. ID	Min. ID	Max. ID	Min. ID
	5.524	5.132	6.135	5.675	6.538	5.92
Base pipe OD (in.)	2.875	2.375	3.50	2.875	4.00	3.50
Screen Wire-Wrap OD (in.)	3.48	2.97	4.13	3.48	4.5	4.13
Clearance (in.)	1.02	1.08	1.00	1.10	1.02	0.90
Casing OD (in.)	7–5/8		8–5/8		9–5/8	
	Max. ID	Min. ID	Max. ID	Min. ID	Max. ID	Min. ID
	7.125	6.435	8.097	7.511	9.063	8.125
Base pipe OD (in.)	4.50	4.00	5.50	5.00	6.625	5.50
Screen Wire-Wrap OD (in.)	5.12	4.50	6.19	5.64	7.23	6.19
Clearance (in.)	1.00	0.97	0.96	0.95	0.92	0.97

3.7.12.4 Gravel-pack job calculations
3.7.12.4.1 Pack-sand volume required

The next stage in the design of the gravel pack is determining the amount of pack-sand required for the treatment. Generally, a worksheet containing information about the geometry of the well is completed (Table 3.7).

This worksheet focuses specifically on the spaces that will be occupied by pack-sand. The standard supplier issue for gravel-pack sand is in 100-lb units (one sack) and the standard sack is considered to occupy $1 ft^3$. Therefore, all of the worksheet information must be converted to cubic feet, as shown in the following equations.

Step 1—Sand Volume Required to Fill Casing/Screen Annulus (Figure 3.9)

$$V_{sa} = (L_s + h) \times V_a \quad (3.1)$$

3.7 Mechanical Method: Techniques and Procedures

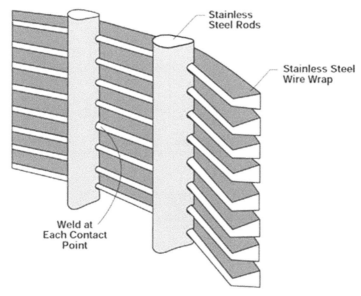

Figure 3.8 Detail of Well Screen Wires Relative to the Vertical Rods.

Table 3.7 Job Calculation Worksheet

1) Depth of the top of the sump packer	5) Inside diameter of casing
2) Length of O-ring sub	6) Outside diameter of screen
3) Length of Tell-tale screen	7) Length of rathole
4) Length of sump packer seal	8) Void volume outside of perforations

where V_{sa} is the sand volume required to fill the annulus, L_s is the combined screen length, h is the height of sand above the screen (with a volume surplus of 40–60 ft), and V_a is a volume factor (ft³/ft).

Step 2—Additional Sand Volume Required to Fill the Rathole (When no Sump Packer is Used) (Figure 3.10)

$$V_{sr} = L_r \times V_c \tag{3.2}$$

where V_{sr} is the sand volume required to fill the rathole, L_r is the rathole length, and V_c is the volume factor for the casing of hole.

Step 3—Sand Volume Required for Outside the Wellbore, Perforation Tunnels and the Formation (Figure 3.11)

$$V_{sf} = L_p \times F \tag{3.3}$$

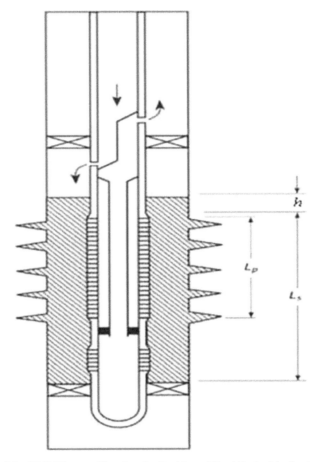

Figure 3.9 Well Diagram Showing the Position of Sand Packed in the Annulus.

where V_{sf} is the sand volume for the formation, L_p is the length of the perforated interval, and F is the volume factor for sand outside the perforations (usually 1 ft³/ft).

Step 4—Total Sand Required (Figure 3.12)

$$V_{st} = V_{sa} + V_{sr} + V_{sf} \qquad (3.4)$$

where V_{st} is total sand volume. For the calculation of total sand weight,

$$W_{st} = V_{st}\rho \qquad (3.5)$$

where W_{st} is total sand weight and is the density factor (100 lb/ft³).

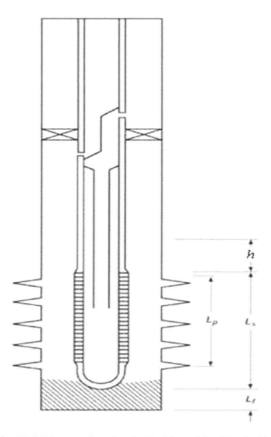

Figure 3.10 Well Diagram Showing the Position of Sand Packed in the Rathole.

3.7.12.4.2 Carrier-fluid Volume

The volume of carrier fluid required to place 10 ft^3 (1000 lb) of sand has been calculated for various sand concentrations per gallon of fluid. Ten cubic feet of sand occupies a real volume of 6.5 ft^3 (48.62 gal) based on specific gravity of 2.63 (for true all-silica sand) and an absolute volume of 0.0456 gal/lb. Table 3.8 shows the carrier-fluid requirements and slurry volumes for multiples and partial increments of 10 ft^3 of sand.

When carrier-fluid volumes are determined for a job, additional fluid is usually set aside for use as a prepad ahead of the sand slurry and as a push pad following the slurry. The recommended volume of additional fluid varies highly throughout the industry, but 6–10 bbl is common for most slurry-packing operations.

110 Sand Control

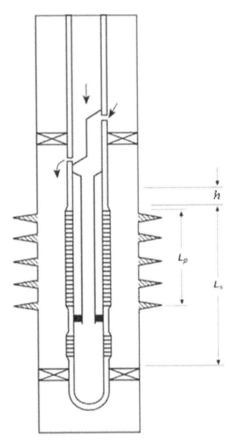

Figure 3.11 Well Diagram Showing the Position of Sand Packed in the Perforation Tunnels.

3.7.12.5 Predicting job outcome by computer modeling

Sand-control engineers can use a variety of versatile computer programs to evaluate different gravel-pack designs. These programs allow engineers to simulate a gravel-pack treatment and track the treatment's progress from initial placement through screenout (Figure 3.13).

Most numerical simulators model the flow of slurry in the wellbore during a gravel pack. Each fluid stage, from the surface to the downhole, is assigned its own fluid rheological properties and solids loading. The pumping can be in the upper circulation, lower circulation or squeeze mode.

Model input screens prompt the user to include detailed parameters from the reservoir, the fluids/slurry anticipated, the packer and tools and well

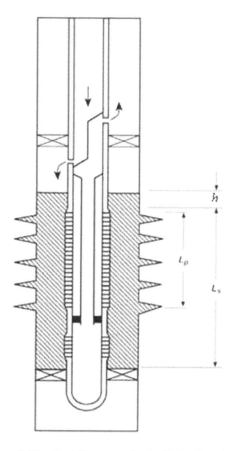

Figure 3.12 Sand Placement for the Entire Gravel Pack.

geometry. These research parameters must be thoroughly researched and properly entered before the model can perform accurate and realistic job predictions.

3.8 Chemical Consolidation Techniques

One of the most compelling reasons to consider chemical consolidation rather than other types of sand exclusion is that allows the wellbore to be free of tools, screens and pack-sand. In certain instances, chemical consolidations can also be performed on the hole without a rig.

Table 3.8 Carrier Fluid Requirement & Slurry Volume to Place 10 ft^3 Sand

Sand per Gallon of Fluid (lb)	Total Slurry Volume (gal)	Volume (gal) Volume (gal)
1.00	1122.91	1074.29
2.00	585.77	537.14
3.00	406.72	358.10
4.00	317.20	268.57
5.00	263.48	214.87
6.00	227.67	179.05
7.00	202.09	153.47
8.00	182.91	134.29
9.00	167.99	119.36
10.00	156.05	107.43
11.00	146.29	97.66
12.00	138.15	89.52
13.00	131.26	82.64
14.00	125.36	76.74
15.00	120.24	71.62

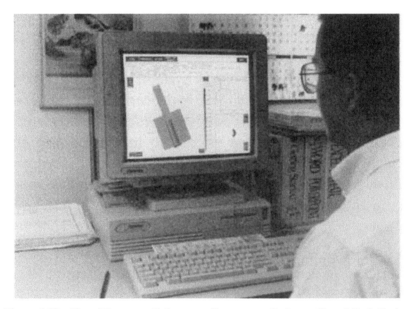

Figure 3.13 Use of Computer Software to Generate an Optimum Gravel-Pack Design.

Sand exclusion by chemical consolidation involves the process of injecting chemicals into naturally unconsolidated formations as a means of providing grain-to-grain cementation. For formation sand consolidation to occur

3.8 Chemical Consolidation Techniques   113

Figure 3.14 Sand Grains Locked Together by In-Situ Resin Consolidation, Leaving Pore Spaces Open to Flow.

with just the contact points of the grain cemented (which creates a continuous flow matrix), excess resin material must be displaced from pore spaces by an overflush fluid at some time in the sequential treatment process. Techniques for successfully accomplishing chemical consolidation are some of the most sophisticated in completion work.

Chemical sand consolidation treatments are formulated to coat individual particles of sand and lock them in place without significantly sacrificing permeability. With the sand consolidated into a hard, permeable mass around the perforations, sand production is minimized and hydrocarbon production is possible for many years (Figure 3.14).

Before a chemical consolidation can be seriously considered, the treatment zone must meet all necessary criteria that will allow the systems to function successfully (Table 3.1). Although exceptions exist, the following criteria must be met:

1. Zone length must generally be no more than 25 ft so that resin and hardener chemicals can be accurately directed to the targeted area.

2. Zone temperature should not exceed 280 °F so that chemicals can be properly placed.
3. Formation permeability should be equal to or greater than 100 md or higher, with less than 15% presence of clays and feldspars.
4. The formation must contain less than 5% calcareous material.
5. The zone must be cased, correctly cemented, and perforated.

The long-term durability of the resin treatment for a chemical consolidation varies. If the previously listed conditions are met to an optimum degree, sand-free service has been documented for as long as 10 years; in other situations, however, sand production may return in only a few months.

The treatment's degree of success depends on the ultimate well conditions imposed and the exact method of consolidation selected (Resvold, 1982). Because of the inherent difficulties associated with chemical consolidation methods, they have a lower success rate than gravel packs. Most failures (either short life or incomplete sand stoppage) can be traced to conditions that exceed the limits of the chemical system used.

1. Internally Activated Systems
2. Externally Activated Systems
3. Application.

3.8.1 Internally Activated Systems

Internally activated systems consist of an epoxy resin that contains a hardener and accelerator. Most of these systems are very time-dependent and react rapidly at high temperatures, resulting in a limited amount of working time. The hardening of the resin begins at the surface. These systems have limited success in very hot wells or in zones that have high clay content. A major internally activated system ensures that all resin placed is mixed with hardener. Under most conditions, these systems demonstrate better-than-average durability.

3.8.2 Externally Activated Systems

Externally activated or overflush systems, as they are sometimes called, use a high-yield furan resin solution. Permeability is established when a specific volume of a spacer is pumped into the formation. This spacer displaces all but a residual resin coating at the grain-to-grain contact points. Afterward, an overflush hardener solution containing an extremely reactive acid component is pumped. A surfactant, which helps the resin adhere to the sand while

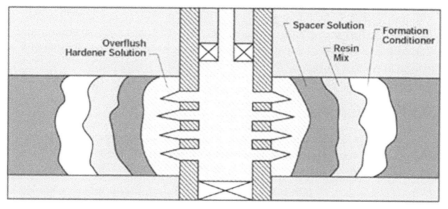

Figure 3.15 Application of In-Situ Resin Consolidation Steps for Internally Hardened System.

extracting the hardener solution's reactive component, also causes the resin to polymerize (Figure 3.15).

Externally activated epoxy systems are also available. Overflushes can be either hydrocarbon or aqueous. Normally, overflushes contain a hardener or accelerator chemical, and some are viscous to improve sweep efficiency. In cases where a two-step overflush is used, one flush restores permeability and the second introduces a cure activator.

3.8.3 Application

The amount of resin solution required for a given application is based on establishing a cylinder of consolidated formation that is 4 ft in diameter. Theoretically, formation inconsistencies nearly always interfere with the uniform penetration of the resin solution. For a cylinder of consolidated formation in typical Miocene sands, 80–90 gal of consolidating fluid is required per foot of formation, based on 20%–30% porosity. This cylindrical matrix provides some casing support and minimizes particle migration. The greater the cylinder's theoretical diameter, the lower the flow velocity at the extremity. Therefore, particles are less likely to be transported at lower velocities.

If a consolidation technique is selected for use on an older well that has produced an appreciable quantity of sand, the well should be packed with clean sand before consolidation. When the liquid resin is pumped into the formation through a cavity or loose zone, uniform distribution of liquid resin is difficult if not impossible. Restressing the formation to some degree is also desirable.

Even if a well or zone meets all the qualifying parameters, other possible objections to performing chemical consolidation include (1) environmental considerations, (2) logistic and storage problems, (3) lack of experience and (4) the increased cost of resins and treating chemicals.

3.9 Combination Methods

Combination methods are sand exclusion methods that incorporate chemical technology, usually resins, to enhance conventional gravel- packing, with or without screens. Two types of resin applications are available for combination treatments: (1) semicured resin coatings can be used on seemingly dry sands with proven latent reactivity and (2) highly reactive, liquid resins can be mixed with hardeners and applied to the pack grains just before the treatment is pumped into the well.

1. Semicured Resin-Coated Pack Gravels
2. Liquid Resin-Coated Pack Gravel

3.9.1 Semicured Resin-coated Pack Gravels

Semicured, resin-coated gravel products are characteristically high- purity, round, crystalline silica sand or bauxite. The base mineral grains are coated with a heat-reactive phenolic resin formulation that contains additives that promote uniform behavior under temperature, fluid and closure stress. The resin coating accounts for 2–4% of the products' weight. The resin is formulated to securely bond to the sand-grain surface. A typical 20/40-mesh round sand has a resin layer less than 0.001 in. thick. The product is dry to the touch and remains chemically active when sufficient heat is applied (Sinclair and Graham, 1977). This condition is referred to as a 'B' stage cure. The ultimate and final cure occurs when temperatures above 130 °F are applied for sustained periods. When bottom hole temperatures are low, heated oil or water can be circulated to further warm the resin-coated particles and help them bond together. If heated fluids are not used, bonding chemicals should be used. These chemicals consist of alcohols and surfactants that soften the resin coating and promote self-adhesion.

In the deliverable form, most curable resin-coated gravel products meet API RP58 specifications and are delivered to the well site in 100- lb bags. A variety of mesh sizes are offered.

The resin coating is tough and slightly deformable, which helps prevent crushing, embedding and flow back after treatments. These resin-coated particles can bond together behind screens or casing with heat or chemical aids. After placement, the bonded, resin-coated gravel is locked into place for long periods. The phenolic resin base used in semi cured, resin-coated gravel materials provides excellent resistance to HCl and HF attack. Future acid cleanout treatments should have little effect on the coating's integrity or strength. The coating helps protect the sand from hot water dissolution to temperatures as high as 600 °F.

Curable, resin-coated gravel products are versatile and can be added as regular gravel to the water or gelled fluid; they can be placed with conventional equipment in a linerless gravel pack or behind a screen or slotted liner. However, since little closure stress is on the resin- coated gravel in these situations, maximum possible strength is seldom attained. Extended shut-in times must often be used so that bond strength can increase.

When correctly placed behind a screen or liner, resin-coated gravel can form a permanent downhole filter, preventing the pack from shifting or mixing with formation fines when the well is opened or shut- in. To a limited extent, resin-coated gravel is used outside cased gravel packs in high-rate wells and/or for long intervals.

Under some conditions, RCS has successfully patched existing gravel-pack screens and liners. During a successful repair, resin- coated particles are injected into the holes of a failed screen or liner. After the particles bond, those remaining in the screen are removed, leaving the hole repaired.

These semicured, resin-coated products are commonly used in the manufacture of prepacked screens and liners. Once tight grain-to- grain contact is attained in the tool, optimum heat curing is applied under controlled conditions. As a result, instead of being loose and mobile, the resin-coated particles within the screen lock together to provide extra protection against particle abrasion and/or erosion.

3.9.2 Liquid Resin-coated Pack Gravel

Procedures to resin-coat pack-sands and pack gravels with liquid resins have been successfully applied and commercially available for several years. These processes differ significantly from ones that feature semicured resin coatings. The liquid resin solution is job-mixed and added to the sand-carrier fluid slurry. Variations of these processes use either an internal or an external resin hardener system. Through processes of chemical attraction, liquid resin is preferentially drawn to the silica sand surfaces instead of well tubular goods.

For the internally hardened variations, hardeners, adhesion promoters and coating aids are added to the liquid resin at the surface. The base resin for these systems is predominantly epoxy. If required, the excess resin can be added to the process, which will permit consolidation of the pack-sand as well as some of the adjacent formation.

Externally hardened systems also begin with liquid resin mixtures added to the sand-fluid slurries. Preferential sand coating with resin also occurs in these processes. When the well no longer accepts the resin-coated sand, the sand slurry remaining in the wellbore is circulated out, leaving the hole clean. This cleanout is followed by a spacer and catalyst solution that hardens the resin-coated sand packed into the perforation and beyond.

Performing these treatments is similar to performing a gravel pack, except the pack-sand ultimately becomes resin-consolidated and theoretically, eliminates the need for screening devices. Several variations of placement procedures are available; some can be continuously blended while others must be batch-mixed.

A common procedure, used in perforated-cased holes requires the use of a viscosified carrier fluid and high concentrations of pack-sand. The sand concentration may be as high as 15 lb/gal. The internally hardened resin concentration is usually 1 gal of resin per sack of pack-sand used. During mixing, the resin coats the pack-sand in the slurry. No provision is made for the excess resin to coat any of the formation faces, except at the interface. The sand, which has a tacky resin coating, remains dispersed and suspended in the carrier fluid. The mixture is pumped to the sanding zone in the well (Figure 3.14, 3.15, 3.16).

Because of the high sand concentration, little carrier fluid is lost to the formation. Since the system is internally catalyzed, flushing out the tacky sand remaining in the wellbore is difficult. Therefore, the entire mass is allowed to set up, and the sand remaining in the casing is drilled out before production (Figure 3.17).

A second procedure used in perforated-cased completions is to disperse resin with low concentrations of sand in a thin carrier fluid in a mixing tank. The sand-to-carrier fluids ratio is normally 1/2 lb/gal. The externally hardened resin-to-sand ratio is from 3–5 gal/sk. During mixing, some of the resin automatically coats the pack-sand and the excess resin remains dispersed in the carrier fluid. As the mixture is pumped, the resin-coated sand screens off against the formation and in the perforation tunnels. As the carrier fluid moves into the formation, tight grain-to-grain contact is established in the pack-sand. The excess resin in the carrier fluid can coat some of the formation

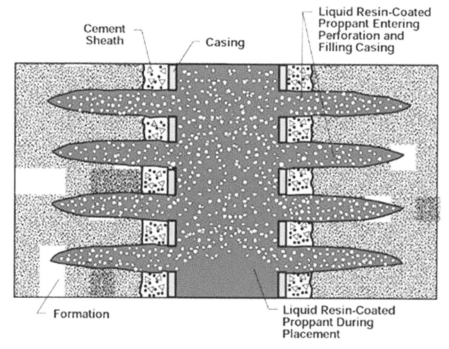

Figure 3.16 Application of Resin-Coated Sand Slurry.

sand. When the well no longer takes any more coated sand, the solids left in the wellbore can be circulated out. The cleanout, in this situation, is followed by a spacer and catalyst solution. This treatment results in a consolidation of both the pack-sand and the formation sand.

3.10 Horizontal Gravel-Packing

Openhole horizontal completions may be extremely difficult to achieve in unconsolidated or weakly consolidated sandstone formations because the horizontal hole is more likely to collapse and fill. Unless certain safeguards are implemented during drilling, the hole will likely collapse before a gravel pack can be performed.

Gravel-packing of high-angle cased wells with long perforated intervals can now be successfully completed sand-free with minimal production loss. Much of the technology used for gravel-packing these highly deviated wells will also likely be used for the successful gravel-packing of true horizontal completions (Elson *et al.*, 1984).

120 *Sand Control*

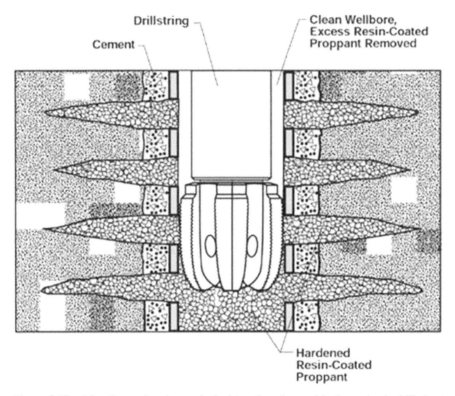

Figure 3.17 After the sand grains are locked together, the sand in the casing is drilled out, leaving the wellbore clear.

Some carrier fluid technology used for vertical wells cannot be successfully used for gravel-packing in high-angle holes. For example, as a gelled carrier fluid breaks back in a vertical wellbore, excess sand in the annular reservoir above the production screen tends to settle and fill any voids in the pack. In a true horizontal gravel pack, the same gravity that corrects a poor gravel pack in a vertical wellbore actually leaves a void across the entire upper side of the annulus. In addition, gravity permits sand that has been packed into the upper perforations to settle back into a loosely packed annulus (Schroeder, 1987; Elson *et al.*, 1984).

1. Variables that Affect Sand Delivery
2. Pump Rate and Fluid Velocity
3. Alpha and Beta Wave Progression Through the Annulus
4. Sand Concentration

5. Placement Procedure and Tool Configuration
6. Liner/Tailpipe Ratio
7. Screen/Casing Clearance
8. Perforation Phasing

3.10.1 Variables that Affect Sand Delivery

According to test results, for a sand-pack without voids, optimum grain-to-grain contact must occur the instant that fluid motion is stopped in a horizontal gravel-packing operation. Therefore, any fluid that does not provide for either complete leak off or complete sand settling could potentially result in a poorly packed annulus.

For gravel-packing horizontal wells, the wellbore deviation is not a variable, nor for all practical purposes are the sand size and density. Therefore, to achieve efficient sand-packing, we are restricted to the variables associated with (1) either the carrier fluid viscosity and density, (2) the sand concentration used and (3) the fluid velocity.

On the basis of field experience and model testing, the following conclusions can be made concerning fluid choice:

- 3.10.1.1 Gelled fluids with 100% sand suspension ability are available, such as xanthan gum gels. However, these gels are generally not good horizontal gravel-pack fluids because of their very low fluid-loss rate to the formation. To ensure a tight pack, the carrier fluid should have an extremely high fluid loss.
- 3.10.1.2 HEC carrier fluids have good sand suspension and better fluid-loss capabilities than xanthan, but they tend to leave voids in the pack because of slow gel-breaking and the retarded development of a tight sand-pack in the annulus.

3.10.2 Pump Rate and Fluid Velocity

The term *pump rate* is often misused as related to gravel-pack operations. Actually, under these circumstances, the pump rate is really referring to the fluid velocity in the annulus or through individual perforations.

The design parameter of pump rate (fluid velocity) directly affects packing efficiency. Tests have shown that in highly deviated wells, a rate increase from 0.6–1.0 bbl/min increased the percentage of packed volume from approximately 60% to approximately 93%. At 1.4 bbl/min, the packed volume increased to approximately 96%. An increase in placement rate virtually eliminates bridges formed in the annulus between the blank pipe and casing above the perforations (Peden et al., 1984).

If we assume that the gel carrier fluid does not have 100% sand suspension capabilities because it was selected to maintain a good fluid loss and allow for quicksand settling, we must determine a compensating pump rate (velocity). The velocity of the carrier fluid through a perforation is affected by two limitations:

- 3.10.2.1 The *maximum velocity* through a perforation must not exceed a rate that would cause a jetting action on the formation, which would intermix the formation sand and pack-sand.
- 3.10.2.2 The *minimum velocity* through a perforation must be high enough to ensure that the pack- sand will flow into the perforation and not past it.

Some studies have shown that viscous carrier fluids provide more efficient perforation packing than brine carrier fluids. This same study also showed that at lower flow rates, the viscous systems tended to pack irregularly, which indicates that higher rates will be beneficial in horizontal gravel packs.

3.10.3 Alpha and Beta Wave Progression Through the Annulus

The basic method for horizontal sand-packing with brine is a two-step procedure, which includes an alpha wave sand deposition in one direction and a beta wave sand deposition in the opposite direction (Dickerson and Anderson, 1987). Water-based sand slurry is pumped down the vertical workstring out the horizontal portion of the screen- casing annulus. The slurry flow rate is controlled so that the slurry velocity is above 7 ft/sec within that tubing. The 7 ft/sec velocity is generally required for the pack-sand to become a slurry.

At the crossover exit into the annulus, the velocity of the slurry rapidly drops to about $1/2$ ft/sec (0.2 m/s) in a normal 4-in. (10 cm) borehole in the formation. The sand immediately falls out of slurry suspensionand a sand dune rapidly builds up in the borehole, both in the forward direction (away from the vertical wellbore) and in the reverse direction (back toward the vertical wellbore). The sand dune fills the horizontal borehole annulus to about 70 to over 90% fill. This deposition is known as the alpha wave sand deposition.

The wedge-shaped form of the leading edge of the deposited dune of sand proceeds at about 1 ft/min (0.5 cm/s) back toward the vertical wellbore. The carrier water fluid splits into two paths within the horizontal borehole. After depositing its sand burden, the main portion of the water progresses through the washpipe and back to the wellbore. The other portion of the brine enters the formation, depending upon its permeability.

The brine velocity in the ullage flow space appears to stabilize automatically. If the water velocity is greater than 1–3 ft/sec (30–91 cm/s), the sand is washed ahead to open a larger ullage flow space. If the water velocity is less than 1–3 ft/sec (30–91 cm/s), the sand deposits more rapidly to fill in and reduce the cross section of the ullage flow space. Based upon many experiments with widely varying conditions of flow, sand concentration and sand size and shape, this sand deposition process appears to be self-regulating.

The actual movement of sand in the ullage space is not caused by the development of slurry. Instead, the sand moves through the hopping (saltation) of individual sand grains along the top of the sand dune, much like the sedimentation process in a river bed.

The leading edge of the sand dune progresses toward the toe of the wellbore until it reaches the end of the screen. At this point, the beta wave deposition of sand in the horizontal borehole is performed. In this beta wave, the sand flows out automatically and simultaneously from the toe of the wellbore back toward the heel end of the horizontal borehole along with the ullage flow space.

The sand movement in the beta deposition occurs in successive waves, during which the sand moves in a wedge-shaped front along the flat top of the sand dune, which had been placed in the alpha wave. After the water propels the sand along the dune top (by saltation), the water then either moves into the formation or is returned by the crossover tool to the surface. The final result is a fully packed horizontal borehole with a screen in place.

3.10.4 Sand Concentration

The design parameter of sand concentration affects the quality of the gravel pack. Gruesbeck *et al.* (1978) illustrated that 'packing efficiency in deviated wellbores increases with (1) lower gravel concentration, (2) decreased particle diameter, (3) decreasing particle density, (4) higher fluid density, (5) higher flow rates and (6) increasing resistance to fluid flow in the tailpipe annulus'.

In most horizontal completions, the leakoff rate should be relatively uniform, since the permeabilities should be relatively uniform. Compensation for some nonuniformity can be accomplished through the use of fluid-loss additives and larger pad volumes ahead of the gravel-pack slurry. These

124 *Sand Control*

fluid-loss additives and larger pad volumes leaked off to the formation can temporarily modify the injection profile, allowing a more uniform slurry loss to the formation while reducing the potential for a premature sand bridge across a higher-permeability interval. Additional compensation can also be accomplished by small reductions in sand concentration and increased pump rates.

3.10.5 Placement Procedure and Tool Configuration

A number of decisions must be made during the design of a horizontal completion system. These decisions affect both the tools that will be run and the operational aspects of running the completion and completing the well.

The job design must be simple to improve reliability and should be integrated so that all components work together efficiently. If the job will be performed offshore, the service provider should preassemble as many tools as possible to accelerate rig floor assembly. If possible, the tools should be run in a lower-cost, less harsh environment so that compatibility can be ensured. Before anything is finalized, the completion tools that extend to the surface should be specified and checked for compatibility. A completion design that does not allow access to tools lower in the completion or a design that restricts flow should be corrected before the design of the horizontal tool system is completed.

Service company personnel should carefully review and be able to complete all aspects of the complete procedures that are developed. These procedures should provide extensive details of all operations so that the job will be performed smoothly without the need for problem-solving and decision-making. Checkpoints, operational warnings, contingency plans for difficult operations and setting depths should all be carefully considered. Fluid-loss control and well control considerations should also be included as part of both the primary procedure and contingency plans so if decisions must be made quickly, they will not severely impact cost or productivity. For offshore operations, the rig-floor assembly should also be discussed as a means of ensuring thorough planning and efficient operation. Future access to the formation to remove fluid-loss devices, bring the well online and perform service operations must also be considered. All of these preparations provide a means of measuring success and improving future well completions.

More complex horizontal tool completion systems allow fluids to be circulated within the wellbore. The primary circulation path is down the workstring and out the shoe at the end of the string. Reverse-circulating is also possible if the system has a shoe that allows circulation in both directions. This shoe must be closed later to prevent the influx of sand. After the packer is set, the circulating paths are commonly alternated from the *heel* (the area below the packer) to the *toe* (the wash shoe at the very end of the completion string). This circulation pattern can be used for displacing fluid-loss materials and drilling solids and/or for stimulating to remove filter cake. It can be followed by a discrete washing of the screen section with cup-packer washing tools.

Once the well has been stimulated and the workstring is removed, a number of mechanical and chemical fluid-loss options exist that can control the well while the uphole completion is run and the wellhead is landed. These options include ceramic flappers, pressure-operated reverse flappers and carbonate fluid-loss materials.

Once the well is controlled, the workstring can be pulled and completion equipment can be run for the uphole completion. Often, the completion equipment is operated just as it would be in conventional vertical well completion. Equipment should be chosen based on the need for future access to the horizontal section and the possibility that coiled tubing may have to be run through flow-control devices.

3.10.6 Liner/Tailpipe Ratio

Numerous studies (Schroeder, 1987; Elson *et al.*, 1984; Peden *et al.*, 1984; Dickerson and Anderson, 1987) have concluded that the liner/tailpipe ratio is probably the most critical aspect of gravel-pack design. For horizontal well completions, this ratio is even more critical. These studies indicate that a ratio of 0.8 or greater is mandatory for the successful gravel-packing of horizontal well completions. The magnitude of the problem and limitations of this chapter does not allow for a more detailed discussion of this very important aspect. However, extensive studies reporting on this critical aspect of gravel-pack design are listed in the reference section.

3.10.7 Screen/Casing Clearance

Studies indicate that the screen/casing clearance directly affects pack efficiency. If the clearance is reduced, velocity will increase, which improves sand transport efficiency. Under these circumstances, however, a greater tendency for premature sand bridging in the screen/casing annulus will occur for two reasons:

- 3.10.7.1 A reduced screen/casing annulus will increase the resistance to flow, which increases flow in the screen- tailpipe annulus.
- 3.10.7.2 Reduced volume in the screen/casing annulus increases the effect of slurry dehydration as fluid leaks off to the formation. Therefore, in horizontal wells, the most desirable means of reducing the tendency for premature sandoff in the screen/casing annulus is to increase the size of the annulus and decrease the sand concentration. Based on the above conclusions, the radial screen/casing clearance for gravel-packing horizontal completions should range from 1.0–1.5 in.

3.10.8 Perforation Phasing

The industry has two seemingly conflicting views on perforation phasing in horizontally cased wells. Some experts recommend perforating only the lower half of the casing or some portion of the bottom half of casing. Others recommend perforating radially around the total casing circumference. Extensive literature is available regarding perforation size, shot density and perforation penetration as they relate to gravel-packing and sand control, but very few references are available regarding perforation phasing as it relates to sand control or gravel-packing. Current practices indicate that horizontal wells should be perforated either 360° or the bottom 120°–180°.

References

Ali, S.A., and Dearing, H.L. (1996). Sand control screens exhibit degrees of plugging. *Pet. Eng. Intl.* 1996, 35–41.

Almond, S.W. and Bland, W.E. (1984). "The Effect of Break Mechanism on Gelling Agent Residue and Flow Impairment in 20/40 Mesh Sand," paper SPE 12485.

API RP-58, second edition, Am. Pet. Inst., Dallas,1995.

Coberly, C.J. and Wagner, E.M. (1938). Some consideration in the selection and installation of gravel-packs for oil wells. *JPT* 1938, 1–20.

Cole, R.C. et al. (1992). "A Study of the Properties, Installation, and Performance of Sintered Metal Gravel-Pack Screens," paper OTC 7012.

Decker, L.R. and Carnes, J.D. (1977). "Current Sand Control Practices," paper IPA, Jarkarta, Indonesia.

Dickerson, W.R. and Anderson, R.R. (1987). "Gravel-packing Horizontal Wells," paper SPE 16931.

Elson, T.D., Darlington, R.H., and Mantooth, M.A. (1984). High-angle gravel-pack completion studies. *JPT* 1984, 69–78.

Gruesbeck, C., Salathiel, W.M., and Echols, E.E. (1978). "Design of Gravel-pack in Deviated Wells," paper SPE 6805.

Hill, K.E. (1941). Factors affecting the use of gravel in oil wells. *Drill. & Prod. Prac. API* 1941, 134–143.

Himes, R.E. (1986). "New Sidewall Core Analysis Technique to Improve Gravel-pack Designs," paper SPE 14813.

Ledlow, L.B. and Johnson, M.H. (1993). "High Pressure Packing with Water: An Alternative Approach to Conventional Gravel- packing," paper SPE 26543.

Maly, G.P. and Krueger, R.F. (1971). Improper formation sampling leads to improper selection of gravel size. *JPT* 1971, 1403–1408.

Maly, G.P. (1979). "Close Attention to the Smallest Job Details Vital for Minimizing Formation Damage," paper SPE 5702.

Morita, N. et al. (1989). Realistic sand-production prediction: numerical approach. *SPEPE* 1989, 11–24; *Trans.*, AIME. **287.**

Mullins, L.D., Baldwin, W.F., and Berry, P.M. (1974). "Surface Flowline Sand Detection," paper SPE 5152.

Murphey, J.R., Bila, V.J., and Totty, K. (1974). "Sand Consolidation Systems Placed with Water," paper SPE 5031.

Peden, J.M., Russel, J., and Oyeneyin, M.B. (1984). "The Design and Optimization of Gravel-packing Operations in Deviated Wells," paper SPE 12997.

Penberthy, W.L. and Shaughnessy, C.M. (1992). "Sand Control," SPE Series on Special Topics, Vol. 1, SPE.

Penberthy, W.L. Jr. (1988). Gravel placement through perforations and perforation cleaning for gravel-packing. *JPT* 1988, 229–236; Trans., AIME, 285.

Rensvold, R. F. (1978). "Full Scale Gravel-pack Model Studies." *EUR 39*, London, England.

Rensvold, R. F. (1982). "Sand Consolidation Resins, Their Stability in Hot Brines," paper SPE 10653.

Saucier, R.J. (1974). Considerations in gravel-pack design. *JPT* 1974, 205–212; Trans., AIME, 257.

Scheuerman, R.F. (1986). A new look at gravel-pack carrier fluid. *SPEPE* 1986, 9–16.

Schroeder, D.E. Jr. (1987). "Gravel-pack Studies in a Full-Scale, High-Pressure Wellbore Model," paper SPE 16890.

Sinclair, A.R. and Graham, J.W. (1977). Super sand control—with resin coated gravel. *Oil & Gas J.* 1977, 56–60.

Sparlin, D.D. (1969). "Fight Sand with Sand—a Realistic Approach to Gravel-packing," paper SPE 2649.

Suman, G.O., Ellis, R.C., and Snyder, R.E. (1983). *Sand Control Handbook*, second edition. Houston, TX: Gulf Publishing Co., pp. 67–68.

Torrest, R.S. (1982). "The Flow of Viscous Polymer Solutions for Gravel-packing Through Porous Media," paper SPE 11010.

van Poolen H.J., Tinsley, J.M., and Saunders, C.D. (1958). Hydraulic fracturing—fracture flow capacity vs. well productivity. *Trans. AIME* 213, 91–95.

Index

A
Aerating Conventional Fluid 12, 19
Alpha and Beta wave progression through the annulus 120, 122
Applications 7, 11, 56
Artificial barriers 43
Atomized Acid 11, 16, 17

B
Basic Equipment 5
Bottom water 36, 49, 57
Bullhead pressure packs 92, 94

C
Case Histories 66, 78
Causes and effects of sand production 75
Causes of sand production 75
Characterizing the problem 51
Chemical consolidation techniques 111
Circulating pressure packs 94
Circulation packs 92, 93
Combination methods 78, 81, 116
Completions-related mechanisms 37
Creative water management 49
Cryogenics 3, 4

D
Description of previously applied treatments 56

Detection and prediction of sand production 77
Displacement 11, 12
Dual completions 43

E
Effects of sand production 76
Expected treatment effect on water production 52
Externally activated systems 114

F
Fines migration 70, 71
Fire Control 12, 24
Foam Cleanout 12, 22
Foam stability and viscosity 23
Foamed Acid 17, 19, 24
Formation characteristics 81, 98
Formation damage 17, 69, 127
Formation Sand 69, 72, 79, 122
Formation sand description 74
Fracpacks 82, 90, 93, 97
Fracturing 19, 43, 46

G
General information 8
Geological deposition of sand 73
Gravel pack job calculations 106
Gravity Pack 92, 93

H

History 33, 34, 37, 81
Horizontal gravel packing 119, 121
Horizontal wells to prevent coning 43

I

In-situ chemical consolidation methods 79
Internally activated systems 114
Introduction 1, 22, 31
Liner ratio 125
Liquid resin coated pack gravel 116, 117

M

Mechanical components 83
Mechanical Job Designs 98
Mechanical method – Techniques and procedures 92
Mechanical methods 78, 81
Mechanical methods of sand exclusion 80
Methods for monitoring and detection of sand production 77
Methods of sand exclusion 80
Methods to predict, prevent, delay and reduce excessive water production 32

N

Nitrified Fluids 11, 15
Nitrogen Service Applications 11

O

Oxygen deficient atmospheres 8, 10

P

Pack-sand selection criteria 100
Perforating 42, 45, 70, 91
Perforating phasing 121, 126
Petro physical properties 73
Pipeline Purging 12, 20
Placement procedures and tool configuration 121, 124
Placement techniques 51, 63, 94, 98
Predicting job outcome by computer modeling 110
Preventing casing leaks 41
Preventing channeling through high permeability 44, 46
Preventing channels behind casing 41
Preventing coning and cresting 42
Preventing excess water production 41, 52
Production restriction 78
Pump rate and fluid viscosity 120, 121
Pumping system 5, 7
Reverse circulation pack 92, 94

S

Safety Bulletin from CGA (Compressed gas association) 9
Sand concentration 82, 109, 120, 123
Sand production mechanisms 71
Screen selection criteria 102
Screen/casing clearance 121, 125
Selecting the appropriate sand exclusion method 80

Selecting treatment composition and volume 62, 63
Semicured resin coated pack gravels 116
Service tools 82, 89, 91
Slurry packs 92, 95, 96, 105
Staged prepacks and acid prepacks 93, 96
Storage tank 7, 76
Summary 97

T
Temperature considerations 64
Tools and accessories 89
Treatment design 47, 51, 52
Treatment lifetime 60
Treatment type 53, 62
Treatment used to reduce excessive water 51

U
Use of foam as a drilling and workover fluid 12, 20
Vaporizer system 2, 7, 8
Variables that affect sand delivery 120, 121
Viscosity considerations 64

W
Washdown method 92, 93
Water control technique 12, 24
Water packs and high rate water packs 93, 96
Water Production Mechanisms 35, 37, 52

About the Author

Mohammed Ismail Iqbal is currently working as a faculty member at the University of Technology and Applied Sciences in Oman and has approximately 11 years of experience nationally and internationally in both teaching and academia. He is pursuing a PhD from Lincoln University College in Malaysia, and has finished a Master's in Oil and Gas Engineering from the University of Salford, United Kingdom, an MBA in Oil and Gas Management from Petroleum University, and a Bachelors in Mechanical Engineering from Osmania University.

Mr. Iqbal has strong corporate relations and industry links, and has designed courses for UVM (Mexico), providing training to big IT giants like ITC Infotech, L&T IES, CAIRN, HCL, PETROSERV, BAPEX, setting up a drilling rig on the UPES campus, which is the first university in India. He has expertise in quality, accreditation, institutional planning (strategic planning, operational planning), quality assessment and experience in launching new programs in digital transformation age of Industry 4.0.

His research interests are production optimization techniques using software like PIPESIM, enhancing recovery well productivity using well testing software like PANSYS, kappa and reservoir simulation using software named RFD, drilling optimization, and development of new chemical for sand control treatment, to name a few.

He has published research papers in Elsevier, Scopus Journal and peer reviewed journals in the research area mentioned above nationally and internationally, and also holds copyright for a research idea (risk assessment) and very recently filed a patent on classroom monitoring during social distancing.